T Cell Protocols

METHODS IN MOLECULAR BIOLOGY™

John M. Walker, Series Editor

514. T Cell Protocols: *Second Edition,* edited by *Gennaro De Libero, 2009*
502. Bacteriophages: *Methods and Protocols, Volume 2: Molecular and Applied Aspects,* edited by *Martha R. J. Clokie and Andrew M. Kropinski, 2009*
501. Bacteriophages: *Methods and Protocols, Volume 1: Isolation,* Characterization, and Interactions, edited by *Martha R. J. Clokie and Andrew M. Kropinski, 2009*
496. DNA and RNA Profiling in Human Blood: *Methods and Protocols,* edited by *Peter Bugert, 2009*
493. Auditory and Vestibular Research: *Methods and Protocols,* edited by *Bernd Sokolowski, 2009*
490. Protein Structures, Stability, and Interactions, edited by *John W. Schriver, 2009*
489. Dynamic Brain Imaging: *Methods and Protocols,* edited by *Fahmeed Hyder, 2009*
485. HIV Protocols: *Methods and Protocols,* edited by *Vinayaka R. Prasad and Ganjam V. Kalpana, 2009*
484. Functional Proteomics: *Methods and Protocols,* edited by *Julie D. Thompson, Christine Schaeffer-Reiss, and Marius Ueffing, 2008*
483. Recombinant Proteins From Plants: *Methods and Protocols,* edited by *Lóic Faye and Veronique Gomord, 2008*
482. Stem Cells in Regenerative Medicine: *Methods and Protocols,* edited by *Julie Audet and William L. Stanford, 2008*
481. Hepatocyte Transplantation: *Methods and Protocols,* edited by *Anil Dhawan and Robin D. Hughes, 2008*
480. Macromolecular Drug Delivery: *Methods and Protocols,* edited by *Mattias Belting, 2008*
479. Plant Signal Transduction: *Methods and Protocols,* edited by *Thomas Pfannschmidt, 2008*
478. Transgenic Wheat, Barley and Oats: *Production and Characterization Protocols,* edited by *Huw D. Jones and Peter R. Shewry, 2008*
477. Advanced Protocols in Oxidative Stress I, edited by *Donald Armstrong, 2008*
476. Redox-Mediated Signal Transduction: *Methods and Protocols,* edited by *John T. Hancock, 2008*
475. Cell Fusion: *Overviews and Methods,* edited by *Elizabeth H. Chen, 2008*
474. Nanostructure Design: *Methods and Protocols,* edited by *Ehud Gazit and Ruth Nussinov, 2008*
473. Clinical Epidemiology: *Practice and Methods,* edited by *Patrick Parfrey and Brendon Barrett, 2008*
472. Cancer Epidemiology, Volume 2: *Modifiable Factors,* edited by *Mukesh Verma, 2008*
471. Cancer Epidemiology, Volume 1: *Host Susceptibility Factors,* edited by *Mukesh Verma, 2008*
470. Host-Pathogen Interactions: *Methods and Protocols,* edited by *Steffen Rupp and Kai Sohn, 2008*
469. Wnt Signaling, Volume 2: *Pathway Models,* edited by *Elizabeth Vincan, 2008*
468. Wnt Signaling, Volume 1: *Pathway Methods and Mammalian Models,* edited by *Elizabeth Vincan, 2008*
467. Angiogenesis Protocols: *Second Edition,* edited by *Stewart Martin and Cliff Murray, 2008*
466. Kidney Research: *Experimental Protocols,* edited by *Tim D. Hewitson and Gavin J. Becker, 2008*
465. Mycobacteria, Second Edition, edited by *Tanya Parish and Amanda Claire Brown, 2008*
464. The Nucleus, Volume 2: *Physical Properties and Imaging Methods,* edited by *Ronald Hancock, 2008*
463. The Nucleus, Volume 1: *Nuclei and Subnuclear Components,* edited by *Ronald Hancock, 2008*
462. Lipid Signaling Protocols, edited by *Banafshe Larijani, Rudiger Woscholski, and Colin A. Rosser, 2008*
461. Molecular Embryology: *Methods and Protocols, Second Edition,* edited by *Paul Sharpe and Ivor Mason, 2008*
460. Essential Concepts in Toxicogenomics, edited by *Donna L. Mendrick and William B. Mattes, 2008*
459. Prion Protein Protocols, edited by *Andrew F. Hill, 2008*
458. Artificial Neural Networks: *Methods and Applications,* edited by *David S. Livingstone, 2008*
457. Membrane Trafficking, edited by *Ales Vancura, 2008*
456. Adipose Tissue Protocols, Second Edition, edited by *Kaiping Yang, 2008*
455. Osteoporosis, edited by *Jennifer J. Westendorf, 2008*
454. SARS- and Other Coronaviruses: *Laboratory Protocols,* edited by *Dave Cavanagh, 2008*
453. Bioinformatics, Volume 2: *Structure, Function, and Applications,* edited by *Jonathan M. Keith, 2008*
452. Bioinformatics, Volume 1: *Data, Sequence Analysis, and Evolution,* edited by *Jonathan M. Keith, 2008*
451. Plant Virology Protocols: *From Viral Sequence to Protein Function,* edited by *Gary Foster, Elisabeth Johansen, Yiguo Hong, and Peter Nagy, 2008*
450. Germline Stem Cells, edited by *Steven X. Hou and Shree Ram Singh, 2008*
449. Mesenchymal Stem Cells: *Methods and Protocols,* edited by *Darwin J. Prockop, Douglas G. Phinney, and Bruce A. Brunnell, 2008*
448. Pharmacogenomics in Drug Discovery and Development, edited by *Qing Yan, 2008*
447. Alcohol: *Methods and Protocols,* edited by *Laura E. Nagy, 2008*
446. Post-translational Modifications of Proteins: *Tools for Functional Proteomics, Second Edition,* edited by *Christoph Kannicht, 2008*
445. Autophagosome and Phagosome, edited by *Vojo Deretic, 2008*
444. Prenatal Diagnosis, edited by *Sinhue Hahn and Laird G. Jackson, 2008*
443. Molecular Modeling of Proteins, edited by *Andreas Kukol, 2008*

METHODS IN MOLECULAR BIOLOGY™

T Cell Protocols

Second Edition

Edited by

Gennaro De Libero

University of Basel, Basel, Switzerland

Humana Press

Editor
Gennaro De Libero
University of Basel
Basel, Switzerland
Gennaro.DeLibero@unibas.ch

Series Editor
John M. Walker
School of Life Sciences
University of Hertfordshire
Hatfield, Hertfordshire, AL10 9AB, UK

ISSN: 1064-3745 e-ISSN: 1940-6029
ISBN: 978-1-58829-587-3 e-ISBN: 978-1-60327-527-9
DOI 10.1007/978-1-60327-527-9

Library of Congress Control Number: 2008941671

Printed on acid-free paper

springer.com

Preface

This book is a collection of protocols, to provide novel techniques for the study of the biology of T lymphocytes.

The methods described in this book do not cover all of the techniques currently used to study T cell-mediated immune responses for the simple reason that T cell immunology is probably the immunological discipline which can be investigated with the widest variety of approaches.

The choice of chapters was made taking into account two points: First, many of the techniques that have been used for some time have been upgraded during the past few years given the greater availability of a variety of products (i.e. cytokines, chemokines, monoclonal antibodies), of refined technical devices (i.e. novel cell culture and cell analysis equipments), and the development of novel instrumentation (i.e. multiparametric flow cytometers, confocal microscopes). Therefore, in several chapters "old techniques", which remain fundamental to T cell immunology, are described in their "modern" versions.

Secondly, the technical advancement has generated the possibility to establish novel assays to investigate T cell physiology. This is reflected in the chapters which describe the protocols that allow use of these modern approaches.

The preparation of this book has required participation of several scientists, all leading experts in their respective fields. Without their enthusiastic participation, this work would not have been possible. Therefore, I thank all authors for their contributions and accept all criticism for missing parts, or information or details, for which only I am responsible.

I hope these protocols will be useful for young investigators who approach for the first time the complex field of immunology and for those more experienced scientists who look for concise and efficacious descriptions of novel methods.

Contents

Preface v
Contributors ix
Color Plates xi

1 Analysis of Frequency and Phenotype of Antigen-Specific T Cells 1
Angus Stock and Vincenzo Cerundolo

2 B Cell Helper Assays 15
Sergio Abrignani, Elena Tonti, Giulia Casorati, and Paolo Dellabona

3 *transkingdom* RNA Interference (*tk*RNAi): A Novel Method to Induce Therapeutic Gene Silencing 27
Thu A. Nguyen and Johannes H. Fruehauf

4 Flow Cytometry and Cell Activation 35
Sonia Gavasso

5 Investigating T Cells by Polychromatic Flow Cytometry 47
Enrico Lugli, Leonarda Troiano, and Andrea Cossarizza

6 Generation of Human T Cell Clones 65
Sabrina Mariotti and Roberto Nisini

7 Limiting Dilution Analysis of Antigen-Specific T Cells 95
Jorge Carneiro, Lurdes Duarte, and Elisabetta Padovan

8 T Cell Epitope-Mapping by Cytokine Gene Expression Assay 107
Maurizio Provenzano and Giulio C. Spagnoli

9 Cytokine Multiplex Immunoassay: Methodology and (Clinical) Applications 119
Wilco de Jager, Berent Prakken, and Ger T. Rijkers

10 Purification of the T Cell Antigen Receptor and Analysis by Blue-Native PAGE 135
Mahima Swamy and Wolfgang W.A. Schamel

11 Non-Replicating Recombinant Vaccinia Virus Expressing CD80 to Enhance T-Cell Stimulation 151
Paul Zajac

Index 163

Contributors

SERGIO ABRIGNANI, MD • *Instituto Nazionale di Genetica Molecolare-INGM, Milan, Italy*
JORGE CARNEIRO, PHD • *Instituto Gulbenkian de Ciência, Oeiras, Portugal*
GIULIA CASORATI, PHD • *Experimental Immunology Unit, Cancer Immunotherapy and Gene Therapy Program, Department of Biology and Biotechnology, San Raffaele Scientific Institute, Milan, Italy*
VINCENZO CERUNDOLO, MD, PHD • *Nuffield Department of Clinical Medicine, Weatherall Institute of Molecular Medicine, Oxford, UK*
ANDREA COSSARIZZA, MD, PHD • *Department of Biomedical Sciences, Section of General Pathology, Modena, Italy*
WILCO DE JAGER, PHD • *Department of Pediatric Immunology, Wilhelmina Children's Hospital, University Medical Center Utrecht, Utrecht, The Netherlands*
GENNARO DE LIBERO, MD, PHD • *Experimental Immunology, Department of Research, University Hospital Basel, Basel, Switzerland*
PAOLO DELLABONA, MD, PHD • *Experimental Immunology Unit, Cancer Immunotherapy and Gene Therapy Program, Department of Biology and Biotechnology, San Raffaele Scientific Institute, Milan, Italy*
LURDES DUARTE, DIPL. BIOL. • *Instituto Gulbenkian de Ciência, Oeiras, Portugal*
JOHANNES H. FRUEHAUF, MD, PHD • *GI Cancer Laboratory, Division of Gastroenterology, Beth Israel Deaconess Medical Center, Harvard Medical School, Boston, MA, USA*
SONIA GAVASSO • *Neurology Research Lab, Haukeland University Hospital, Gamle, Havedbgning, Bergen, Norway*
ENRICO LUGLI, BSC • *Department of Biomedical Sciences, Section of General Pathology, Modena, Italy*
SABRINA MARIOTTI, PHD • *Dipartimento di Malattie Infettive, Parassitarie e Immunomediate, Istituto Superiore di Sanità, Roma, Italy*
THU A. NGUYEN, BSC • *GI Cancer Laboratory, Division of Gastroenterology, Beth Israel Deaconess Medical Center, Harvard Medical School, Boston, MA, USA*
ROBERTO NISINI, MD • *Dipartimento di Malattie Infettive, Parassitarie e Immunomediate, Istituto Superiore di Sanità, Roma, Italy*
ELISABETTA PADOVAN, PROF. PHD • *Universidade de Lisboa, Faculdade de Medicina, Lisboa, Portugal*
BERENT PRAKKEN, PROF. MD, PHD • *Department of Pediatric Immunology, Wilhelmina Children's Hospital, University Medical Center Utrecht, Utrecht, The Netherlands*
MAURIZIO PROVENZANO, MD • *Department of Urology, University Hospital of Zurich, Zurich, Switzerland*

GER T. RIJKERS, PHD • *Department of Pediatric Immunology, Wilhelmina Children's Hospital, University Medical Center Utrecht, Utrecht, The Netherlands; Laboratory of Medical Microbiology and Immunology, St. Antonius Hospital, Nieuwegein, The Netherlands*

WOLFGANG W.A. SCHAMEL, PHD • *Department of Molecular Immunology, Max Planck Institute for Immunobiology, University of Freiburg, Freiburg, Germany*

GIULIO C. SPAGNOLI, MD • *Institute of Surgical Research and Hospital Management, University Hospital Basel, Basel, Switzerland*

ANGUS STOCK, BSC • *Nuffield Department of Clinical Medicine, Weatherall Institute of Molecular Medicine, Oxford, UK*

MAHIMA SWAMY, MSC, ME • *Department of Molecular Immunology, Max Planck Institute for Immunobiology, University of Freiburg, Freiburg, Germany*

ELENA TONTI, PHD • *Experimental Immunology Unit, Cancer Immunotherapy and Gene Therapy Program, Department of Biology and Biotechnology, San Raffaele Scientific Institute, Milan, Italy*

LEONARDA TROIANO, BSC • *Department of Biomedical Sciences, Section of General Pathology, Modena, Italy*

PAUL ZAJAC, PHD • *University Hospital Basel, Institute of Surgical Research and Hospital Management, Basel, Switzerland*

Color Plates

Color Plate 1: Multidimensional analysis of human PBMC stimulated with either IL-6 (*green*), IL-4 (*red*) or *left* untreated (*blue*).
Cells were fixed and permeabilized following protocol 3.2 and stained simultaneously with antibody cocktail A. *Top panels* show superimposed *dot plots* and *histograms* for T-cells ($CD3^+$), the *bottom panels* show B-cells ($CD20^+$). In overlays the induction of specific phosphorylation events are clearly identifiable. (*see* discussion on p. 41)

Color Plate 2: PBMC were stimulated with indicated cytokines, fixed and permeabilized according to protocol 3.2.
T-cells ($CD3^+$) and B-cells ($CD20^+$) were gated according to markers while monocytes were gated in *scatter plot. Open histograms* represent untreated cells, *filled histograms* stimulated cells. Induction of phosphorylation is clearly identifiable (*filled yellow histograms*). (*see* discussion on p. 41)

Color Plate 3: Visualization of the data generated by the FACS analysis following protocol 3.2.
The columns represent the cell subsets, T-cells, B-cells, monocytes. Each row represents a cytokine stimulation stained with one of the antibody cocktails and subsequently analyzed for the indicated phosphoprotein. The color of each block represents the fold change (log_2) in MFI in the channel corresponding to the analyzed phophorylated protein. (*see* discussion on p. 42)

Chapter 1

Analysis of Frequency and Phenotype of Antigen-Specific T Cells

Angus Stock and Vincenzo Cerundolo

Abstract

Over the last decade, our understanding of the cellular immune system has been greatly advanced through the development of methods to identify antigen-specific T cells directly ex vivo. The major reagents and techniques used for this purpose are (i) tetramerised MHC:peptide complexes (tetramers) which bind to specific T-cell receptors (TCR) and (ii) assays that detect T cells which synthesise cytokines in response to cognate stimulation (intracellular cytokine staining (ICS)). Here, we provide a detailed description of the procedure for generating and using class I MHC:peptide tetramers to label peptide-specific T cells and for carrying out ICS to measure antigen-specific T lymphocytes.

Key words: Tetramers, antigen-specific T cells, MHC class I, intracellular cytokine staining, CTL.

1. Introduction

Accurate measurements of MHC class I restricted T cells have been hampered by the lack of staining reagents capable of identifying antigen-specific cytotoxic T lymphocytes (CTL). Until a few years ago, assays used to detect CTL depended on in vitro culture of antigen-specific CTL and relied on the ability of expanded CTL either to kill target cells or to secrete relatively large amounts of lymphokines. Limiting dilution assay (LDA) was used to quantify CTL precursor frequency and Cr^{51} release assay was used to assess CTL specificity. Both assays were unable to detect cells incapable to proliferate or to show cytotoxic effector function. New methods for detection of antigen-specific CTL have recently been developed (i.e. ELISPOT and intracellular cytokine staining (ICS)) and have allowed to measure in ex vivo assays the frequency

Gennaro De Libero (ed.), *T Cell Protocols: Second Edition, vol. 514*

DOI 10.1007/978-1-60327-527-9_1 Springerprotocols.com

of cytokine-secreting effector cells *(1–4)*. Although these assays can reliably measure the frequency of specific CTL able to secrete cytokines, they fail to detect resting or naïve CTL, unable to secrete lymphokines in ex vivo assays. Over the last few years, a novel method has been developed based on the generation of fluorogenic tetrameric HLA class I molecules loaded with defined peptide epitopes, which is independent of the ability of cells to proliferate and secrete lymphokines *(5)*. The use of HLA class I tetramers has allowed the characterisation and monitoring of specific viral and tumour responses and has provided an opportunity to greatly accelerate the development of new vaccination strategies (reviewed in *(6)*). Here we will describe Materials and Methods to engineer MHC class I tetramers and to carry out combined MHC class I tetramer and ICS of antigen-specific T lymphocytes.

2. Materials

2.1. MHC Class I: Peptide Tetramers

1. pET expression systems (Novagen)
2. Oligonucleotide primers, DNA polymerase, ligase and restriction enzymes
3. Agarose and DNA-sequencing systems
4. Ampicillin (used at 100 μg/ml unless otherwise stated)
5. IPTG
6. Triton wash: 50 mM Tris–HCl (pH 8.0), 100 mM NaCl, 0.5% Triton, 10 mM DTT, 1 mM EDTA and 0.1% azide
7. Re-suspension buffer: 50 mM Tris–HCl (pH 8.0), 100 mM NaCl, 10 mM DTT and 1 mM EDTA
8. Urea solution: 8 M urea, 10 mM Tris–HCl (pH 8.0), 100 mM NaH_2PO_4, 0.1 mM EDTA and 10 mM DTT
9. BCA protein assay kit
10. Reagents and instruments for SDS gel
11. 0.45 μM filter membrane
12. Refolding buffer: 100 mM Tris–HCl (pH 8.0), 400 mM l-arginine hydrochloride, 2 mM EDTA, 5 mM reduced glutathione, 0.5 mM oxidised glutathione and 0.1 mM PMSF (diluted in H_2O)
13. Minimal peptide
14. Amicon stir cell (Diaflo PM10 150 mm membrane)
15. FPLC

16. Tris-buffered saline
17. $MgCl_2$(stock solution, 50 mM)
18. Leupeptin (stock solution, 10 μM)
19. Pepstatin (stock solution, 10 μM)
20. d-Biotin (stock solution, 4 mM)
21. ATP (stock solution, 50 μM)
22. TBS
23. Bir A Enzyme (stock solution, 500 μM)
24. Phycoerythrin-conjugated streptavidin
25. ELISA reagents
26. Streptavidin-peroxidase
27. FACS buffer: PBS containing 1% BSA and 0.02% sodium azide
28. Antibodies against T-cell co-receptors (CD8 and CD4)
29. Propidium iodide
30. Formaldehyde

2.2. Intracellular Cytokine Staining

1. Synthetic peptides
2. Brefaldin A (diluted in methanol)
3. Antibodies against surface antigens (e.g. CD8, CD4, etc.) and intracellular cytokines (e.g. IFN-γ, IL-2, TNF-α, IL-4, IL-10, etc.)
4. Formaldehyde: 1% diluted in PBS with 0.1% sodium azide
5. Saponin
6. T-cell medium: RPMI 1640 supplemented with 10% heat-inactivated foetal calf serum and 5% supplementum complementum (SC: HEPES 23.83 g/l, benzylpenicillin 2×10^6U/l, streptomycin 2 g/l, 1-glutamine 6 g/l for RPMI, 50 μM 2-β mercaptoethanol)
7. FACS buffer: PBS containing 1% BSA and 0.02% sodium azide.

3. Methods

Here we shall provide protocols for the two most commonly used flow cytometry-based techniques for enumerating antigen-specific T cells: (1) MHC tetramer analysis and (2) ICS. We shall describe the procedure for generating of class I MHC:peptide tetramers (**Section 3.1**) and using these reagents to stain population of cells for identifying antigen-specific T cells (**Section 3.2**). Finally, the protocol for using ICS to identify both peptide- and virus-specific T cells will be described (**Section 3.3**).

3.1. Production of MHC Class I:Peptide Tetrameric Complexes

The strategy for generating class I MHC:peptide complexes was first described by Garboczi and Wiley in 1992 *(7)*. In this method, the β_2-microglobulin (β_2-m) and the MHC heavy chains are expressed separately in prokaryotic expression systems. These molecules are then purified from bacterial inclusion bodies and refolded in the presence of peptide. However, monomeric complexes bind TCR poorly, necessitating the formation into multimers to allow cooperative TCR binding. The method for combining MHC:peptide monomers into tetrameric complexes for staining antigen-specific T cells was devised by Altman and Davis in 1996 *(5)*. In this protocol, refolded monomers (which are fused to the Bir A substrate peptide) are biotinylated with the BirA enzyme and tetramerised with fluorochrome-conjugated streptavidin. Here, we shall provide an outline for this procedure, although it should be noted that customised tetramers are commercially available and may be purchased from outside sources.

3.1.1. Expression of MHC Chains

The most commonly used vectors to express MHC proteins are the pET plasmids (although other vectors may be used in their place). Genes inserted into the multiple cloning site of the pET plasmids are expressed as a T7 gene product, induced through the addition of IPTG.

1. Amplify the extracellular domains of the MHC class I heavy chains ($\alpha1, \alpha2$, and $\alpha3$) or β_2-m from a cDNA or/and existing plasmid and ligate into separate expression vectors. In the case of the MHC heavy chain, insert sequence immediately upstream of the BirA substrate peptide (BSP) sequence to produce MHC–BSP fusion product.
2. Transform plasmids into *E. coli* and incubate over night at 37°C with ampicillin selection (100 μg/ml). Pick single colonies, reselect with ampicillin and screen for the desired plasmid through enzyme digestion and sequence analysis.

3.1.2. Induction of MHC Protein

1. Using a single colony containing either the heavy chain or β_2-m expression vectors, inoculate 10 ml LB (+ampicillin selection) and grow overnight (37°C, shaking). Use 1 ml of this 'starter' broth to inoculate 'bulk' 200 ml LB (+ampicillin selection) and culture overnight (37°C, shaking).
2. The following day, add 50 ml of the bulk culture to each of four flasks containing 1 l LB (+ampicillin selection). Culture at 37°C with shaking, until the OD_{600} reaches 0.6 (this takes ~3–4 h). Take a 1.5 ml sample to analyse pre-induction protein expression (see **Fig. 1.1**).

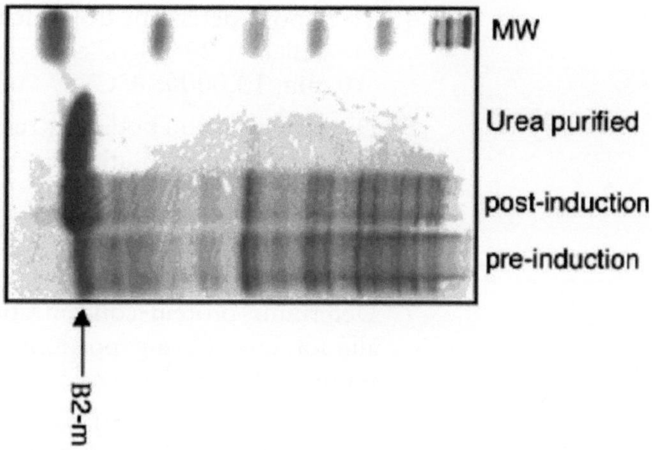

Fig. 1.1. Purification of MHC proteins.
Protein extracts from bacteria transformed with β_2-m expression vectors taken prior to IPTG induction (pre-induction), 4 h after IPTG addition (post-induction) and following urea purification were resolved on a 15% SDS-PAGE gel and stained with Coomassie blue. Note the induction and purification of the β_2-m sized band.

3. Induce the expression of MHC proteins by adding IPTG (final concentration 0.5 mM) and incubate for 4 h (37°C, shaking) to allow MHC protein expression.
4. Take a 1.5 ml post-induction sample for gel analysis (see **Fig. 1.1**), measuring the OD_{600} to equilibrate total pre- and post-induction bacterial concentration.
5. Pellet the remaining culture (20 min, 5000 *g*, 4°C). Decant supernatant and resuspend in ice-cold PBS (~10 ml per flask). Here, bacteria may be frozen at –80°C and stored overnight before inclusion body purification, or purified directly.

3.1.3. Inclusion Body Isolation

1. If frozen, thaw pellet on ice. Lyse bacteria by sonicating in bursts of 30 s, keeping on ice between bursts to prevent overheating. Repeat until suspension reaches milk-like consistency (this normally requires ~4 rounds of sonication).
2. Transfer suspension to centrifuge tubes and spin for 10 min (13,000 *g*, 4°C). Both inclusion bodies (which should appear white) and cell debris (which appear as darkish matter) should pellet. If the supernatant remains cloudy re-spin.
3. To remove cell debris, resuspend cell pellet in 25 ml of cold triton wash buffer. Spin for 10 min (13,000 *g*, 4°C) and discard supernatant. Wash another 2–3 times in triton buffer or until the supernatant has become clear (indicating that bacterial debris has been removed).

4. To remove detergent from inclusion body sample, resuspend the pellet in 25 ml of cold re-suspension buffer and pellet (10 min, 13,000 *g*, 4°C).
5. Dissolve inclusion bodies by resuspending in a total volume of 25 ml fresh urea solution, mixing with a glass homogeniser. Rotate overnight at 4°C.
6. Spin urea extracts for 10 min (13,000 *g*, 4°C). Working on ice, pool supernatant and take a sample for gel analysis (see **Fig. 1.1**).
7. Determine protein concentration (using BCA analysis) and aliquot into 13 mg portions. Store at –80°C for up to 1 month before refolding.
8. For gel analysis, pellet 1 ml of pre-induction and post-induction samples and resuspend in double-distilled water at a volume of 100 µl × OD_{600} (equilibrating protein concentration). Run 10 µl of these resuspended samples alongside the urea-purified protein on a SDS gel (10% gel for MHC heavy chains and 15% for β_2-m: see **Fig. 1.1**).

3.1.4. Refolding and Purification of Class I MHC:Peptide Monomers

The MHC heavy chain and β_2-m proteins are mixed with the target peptide, refolding into monomeric MHC:peptide complexes. Monomers are then purified by chromatography.

1. Make up 1 l of refolding buffer and filter through a 0.45 µm membrane. Collect into a 2 l conical flask and cool to 4°C.
2. Dissolve 10 mg of peptide in 1 ml DMSO immediately before refolding.
3. Stir refolding buffer at 4°C and add dissolved peptide (final concentration ~10 µM), 26 mg β_2-m protein (final concentration 2 µM) and 32 mg MHC heavy chain protein (final concentration 1 µM). Refold for 40 h at 4°C with stirring.
4. To concentrate the refolded protein, filter the mix through a 0.45 µM membrane to remove precipitated proteins and load into an Amicon stir cell with a Diaflo PM10 150 mm membrane. Concentrate to 7 ml volume.
5. Refilter concentrate through a 0.45 µM membrane and purify the monomer using FPLC with Tris-buffered saline as the running buffer. Collect fractions of appropriate molecular weight and concentrate to below 3 ml using an Amicon stir cell (working at 4°C to limit protein degradation).

3.1.5. Biotinylation of Class I MHC:Peptide Monomers

At this stage, monomeric class I MHC:peptides are biotinylated through the addition of the BirA enzyme. This enzyme biotinylates the BSP that is fused to the MHC heavy chain.

1. Transfer the purified monomers to a 50 ml Falcon tube and add
   - $MgCl_2$ (final concentration 5 mM)

- Leupeptin (final concentration 1 μM)
- Pepstatin (final concentration 1 μM)
- d-Biotin (final concentration 400 μM)
- ATP (dissolved in TBS: final concentration 5 mM)
- Bir A enzyme (final concentration 50 μM)

Make up to a final volume of 4 ml with TBS, mix through inversion and incubate overnight at room temperature.

2. Filter the biotinylated complex through a 0.45 μm membrane and purify with FPLC (Tris-buffered saline running buffer). Collect fractions of appropriate molecular weight and immediately add leupeptin and pepstatin (final concentration of each at 1 μM) to prevent protein degradation.
3. Concentrate biotinylated complexes on an Amicon cell to a final volume of around 1 ml.
4. Determine protein concentration (i.e. with a BCA assay) and aliquot into 100 μg portions. Quick-freeze monomers in liquid nitrogen and store at –80°C until tetramerisation.
5. See Notes for optional step (*see* **Note 1**).

3.1.6. Tetramerisation of Class I MHC:Peptide Monomer

Monomers are tetramerised through the addition of fluorochrome-conjugated streptavidin. To achieve maximal tetramerisation, streptavidin is initially added to an excess of monomer. Limiting avidin ensures that all biotin-binding sites are occupied with MHC monomers, promoting the formation of tetrameric complexes. The concentration of avidin is then progressively increased until all MHC has gone into tetramers and the avidin is saturating. Here, ExtrAvidin–Phycoerythrin (Sigma: E-4011) is used at a final concentration of 1 μl/μg of monomer (for instance, a total of 200 μl of avidin–PE to tetramerise 200 μg monomer).

1. Initially add half the total avidin–PE volume to monomer, mixing gently with a pipette tip. Incubate on ice for 30 min (keep in dark to protect fluorochrome).
2. Divide the remaining avidin–PE into five equal volumes.
3. Add one aliquot of avidin–PE (equating to one tenth of total volume) to monomer. Mix and incubate for 20 min on ice (dark).
4. Repeat step 3 another four times until the entire volume of avidin–PE has been added.
5. Test biotin availability following tetramerisation by ELISA. Plate out equivalent concentrations of monomer and tetramer in ELISA plates and perform twofold serial dilution (across eight wells). Perform standard ELISA, probing for biotin using

streptavidin-peroxidase. Upon developing, the tetramer sample should appear to contain 1/8–1/16 as much biotin as the monomer. If biotin levels are in excess of this, repeat steps 3 and 4 until avidin has saturated biotin sites.

3.2. Using Class I MHC:Peptide Tetrameric Complexes for Frequency Analysis of Antigen-Specific T Cells

In this section, we shall describe the staining protocol for using class I MHC:peptide tetramers to identify antigen-specific T cells in whole lymphocyte populations.

3.2.1. Staining Protocol for Class I MHC Tetramers

1. Prepare single lymphocyte suspension from blood or organ tissues of test and control samples (*see* **Note 2**).
2. Lyse red blood cells and count live lymphocytes.
3. Transfer 1–2 $\times 10^6$ live lymphocytes into staining vessels (usually 5 ml FACS tubes or 96-well plates). Pellet cells by centrifugation (5 min, 500 *g*) and wash twice with FACS buffer.
4. Stain cells first with tetramer by resuspending the cell pellet in 50 μl tetramer complex diluted to the optimal working concentration in FACS buffer (*see* **Note 3**). Incubate for 30 min at 37°C in the dark (*see* **Note 4**).
5. Wash cells with FACS buffer and pellet (5 min, 500 *g*).
6. Next stain for T-cell co-receptors. Resuspend cell pellet in a saturating concentration of antibodies (50 μl volume diluted in FACS buffer) against CD8 (for class II tetramers stain for CD4). In addition, antibodies directed against T-cell surface antigens may be included at this step (*see* **Note 5**). Incubate for 30 min on ice in the dark.
7. Wash cells twice in FACS buffer and prepare for flow cytometric analysis. The sample may be analysed directly as live cells or fixed and stored for up to 2 days before analysis.
   a For live analysis, resuspend cell pellet in 200 μl FACS buffer. Just prior to flow cytometry (between 1 and 30 min) add propidium iodide (final concentration at 2.5–5 μg/ml), to allow dead cell exclusion during analysis (*see* **Note 6**).
   b For fixation, resuspend cell pellet in 100 μl of 1% formaldehyde and incubate for 20 min at room temperature (dark). Following fixation, wash cells twice in PBS. Resuspend in 200 μl PBS and store at 4°C in the dark until flow cytometry.

3.2.2. Flow Cytometric Analysis of Tetramer Staining

1. For analysis, set up FSC versus SSC plot and set the primary gate (R1) upon lymphocyte-sized cells.
2. Where propidium iodide (PI) has been included set the secondary gate (R2) upon PI-negative lymphocytes. Combine gates (R1 × R2) to restrict analysis to live lymphocytes (see **Fig. 1.2**).
3. To analyse the frequency of antigen-specific T cells, show tetramer staining versus CD8 expression. Traditionally, the frequency of antigen-specific T cells is expressed as the percentage of T cells (CD4 or CD8) that stain positive with the MHC:peptide tetramer complex. Alternatively,

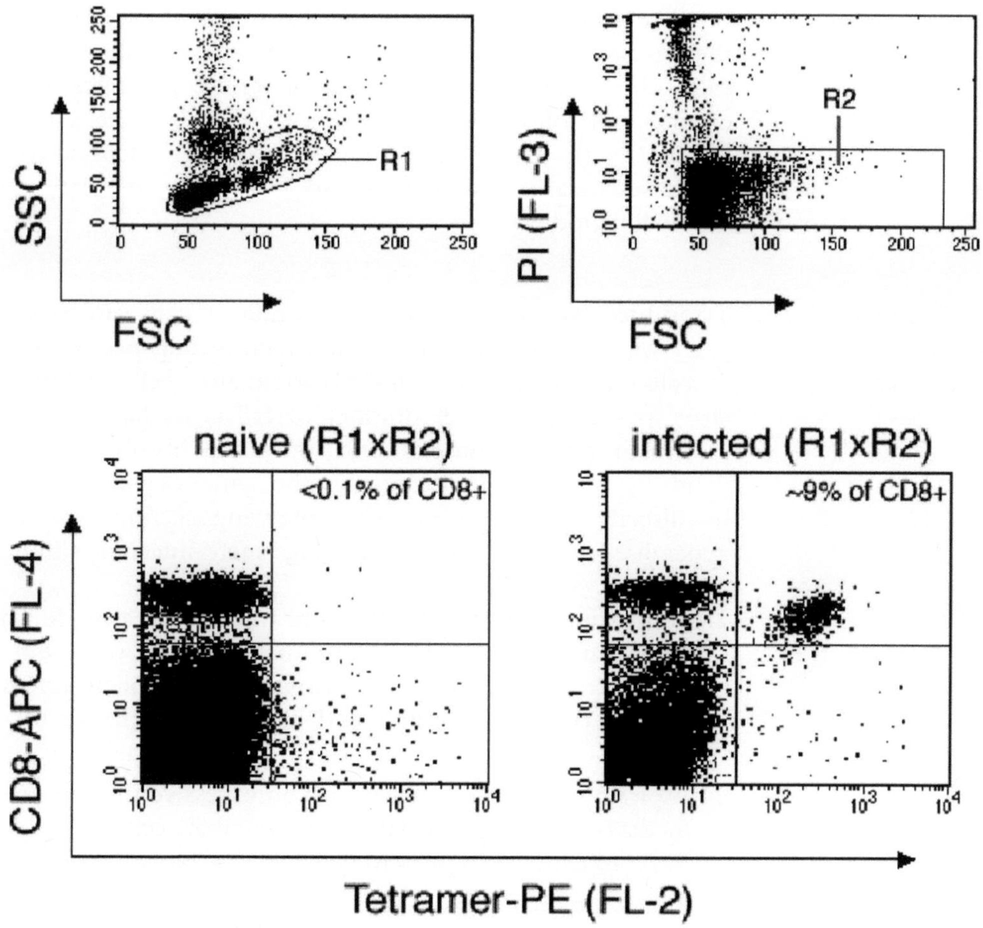

Fig. 1.2. Quantitation of antigen-specific T cells by tetramer staining.
Blood from naïve C57BL/6 mice and mice infected 7 days earlier with herpes simplex virus (HSV) were stained with an APC-conjugated anti-CD8 antibody and the PE-conjugated tetramer of the dominant HSV peptide in complex with H2-K^b. Frequency analysis was restricted to live lymphocytes (R1 × R2) by gating upon lymphocyte-sized events (R1) that were PI negative (R2). *Inset* values show the percentage of CD8+ T cells that are tetramer positive for naïve and infected samples. Note that the HSV tetramer-specific population represents less than 0.1% of CD8+ T cells from naïve mice before expanding to around 9% of circulating CD8+ T cells after HSV infection.

when cells are from organ samples, the total number of antigen-specific T cells per organ may be a more appropriate value (see **Fig. 1.2**).

3.3. Analysis of Antigen-Specific T Cells Using Intracellular Cytokine Staining

Upon recognition of their cognate antigen, T cells synthesise and secrete a range of cytokines, including IFN-γ, TNF-α, IL-2, IL-4 and IL-10. The ICS assay utilises this activity, identifying antigen-specific T cells through their production of cytokines in response to antigen. In this assay, the secretory pathway is blocked, causing the intracellular accumulation of nascent cytokines, which in turn are detected by antibody staining and flow cytometry *(1–4)*.

The primary advantage of the ICS versus tetramer staining is that precise knowledge of T-cell epitopes is not required. Specifically, the T-cell response against an entire pathogen may be enumerated following activation with stimulator cells that are infected with particular pathogen. However, peptide-specific responses are also routinely measured following stimulation with minimal peptides.

3.3.1. Preparation of Virus-Infected Stimulator Cells

To induce cytokine synthesis, responder T cells can be stimulated with either minimal peptides (stimulating peptide-specific T cells) or with virus-infected stimulator cells (stimulating virus-specific T cells). A number of cell types have been used as virus-infected stimulator cells, including fibroblasts *(8)* and dendritic cells *(9)*. Additionally, the lymphocyte sample may be directly infected *(10)*. The following section describes a generalised method for generating such infected stimulator cells.

1. Wash stimulator cells with serum-free media, aspirate supernatant and resuspend in a minimal volume of serum-free media containing the target virus at a multiplicity of infection (m.o.i) of between 1 and 5. Incubate cells for 60 min at 37°C.
2. Aspirate media and replace with an excess volume of growth media (e.g. RPMI with 10% FCS). Incubate cells for 4–16 h at 37°C to allow virus protein synthesis (*see* **Note 7**).
3. Following incubation, aspirate infectious supernatant, harvest cells and resuspend in T-cell media for use as virus-infected stimulators.

3.3.2. *In Vitro* Stimulation of T Cells

1. Prepare single lymphocyte cell suspensions from blood or organ tissue from test and control samples. Lyse red blood cells and wash in T-cell growth media (*see* **Note 2**).

2. Count live lymphocytes and transfer 1–2 $\times 10^6$ live lymphocytes into wells of a 96-well round-bottom plate. In general, place each sample into four wells; duplicate wells with and without antigenic stimulation.
3. Stimulate T cells by adding either the cognate peptide (final concentration of between 0.1 and 1 μM) or virus-infected stimulator cells (between 1 and 2×10^5 cells per well) diluted in T-cell media. For unstimulated controls, add media alone or non-infected stimulator cells.
4. Immediately add Brefeldin A (*see* **Note 8**) at a final concentration of between 5 and 10 μg/ml. Culture cells for 6 h at 37°C in 6.5% CO_2 to allow the intracellular accumulation of newly synthesised cytokines.

3.3.3. Staining and Analysis for ICS Assays

1. Following in vitro stimulation, pellet cells (5 min, 500 *g*, 4°C) and wash twice in FACS buffer.
2. Resuspend the cell pellet in a saturating concentration of anti-CD8 and/or anti-CD4 antibodies and stain on ice for 30 min in the dark. In addition to T-cell co-receptors, it is useful to include antibodies directed against surface antigens that are constitutively expressed upon antigen-experienced T cells such as CD44 and CD11a/b (mouse) or CD45RA (human). Increasing the number of staining parameters enhances specificity and is particularly useful for rare populations.
3. Following surface staining, wash twice with FACS buffer to remove unbound antibody and pellet cells (5 min, 500 *g*, 4°C).
4. Resuspend cell pellet in 100 μl of 1% formaldehyde and stand for 20 min at room temperature in the dark. Following fixation, wash cells twice with PBS.
5. Dilute antibodies against target cytokine(s) (e.g. IFN-γ, TNF-α, IL-2, IL-4 or IL-10) in 0.5% saponin. Pellet fixed cells (5 min, 500 *g*, 4°C) and resuspend in 50 μl of cytokine solution (separately stain stimulated cells with isotype control antibodies to validate cytokine specificity). Incubate cells at 4°C in the dark for at least 1 h (alternatively, cells may be left overnight in antibody suspension).
6. Following incubation, wash cells twice in PBS (this may include 0.5% saponin if background levels are high). Resuspend cell pellet in 200 μl PBS for flow cytometry.
7. For flow cytometric analysis, set the primary gate on lymphocyte-sized cells and analyse the expression of cytokines versus CD4 or CD8 (*see* **Fig. 1.3**).

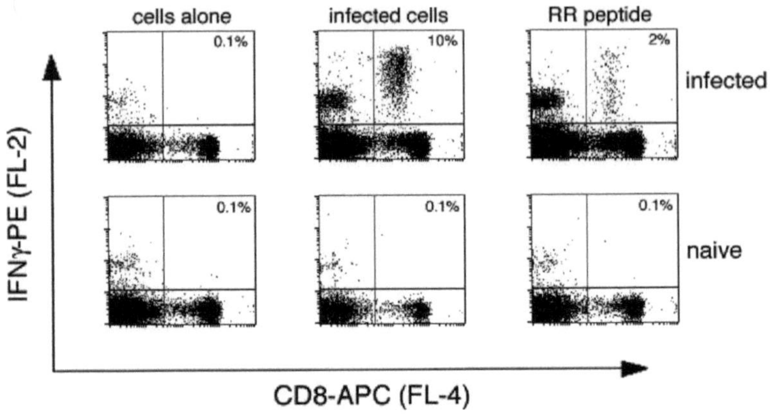

Fig. 1.3. Quantitation of virus and peptide-specific T cells by ICS.
Splenocytes from naïve C57BL/6 mice and mice infected 8 days earlier with HSV were stimulated with non-infected fibroblasts (cells alone), HSV-infected fibroblasts (infected cells) or a single HSV peptide (RR peptide). After 6 h, cells were stained for CD8 and intracellular IFN-γ. Dot-plots are gated upon lymphocyte-sized events and show the percentage of CD8+ T cells that are IFN-γ+ (*inset* value). Note that less than 0.1% of splenic CD8+ T cells from naïve mice produce IFN-γ following stimulation with either HSV or peptide, while around 2 and 10% of CD8+ T cells from HSV-infected mice produce IFN-γ following peptide or HSV-infected cell stimulation, respectively.

4. Notes

1. It may be worthwhile to confirm that the monomer has been successfully biotinylated before commencing tetramerisation. Biotinylation may be confirmed simply through the use of a commercial biotin-ELISA detection kit. However, the sample may appear biotin-positive when only a small fraction of total monomers are actually biotinylated. In this case it is possible to purify biotinylated proteins through MonoQ-ion exchange. In MonoQ-ion exchange, biotinylated monomers run slightly heavier, allowing their separation from lighter/non-biotinylated monomers.
2. To validate the specificity of tetramer and ICS for antigen-specific T cells, it important to assess the background levels of staining against non-specific T cells. To do this, each experiment should include negative control samples that are known to have very few T cells of the defined specificity (<0.1% of total T cells). Suitable controls include individuals that have not been challenged with the vaccine/infection (i.e. 'naïve' controls) or individuals of a distinct MHC haplotype.
3. For each new tetramer reagent (either produced in house or made commercially) it is important to confirm the ability to stain antigen-specific T cells and secondly to determine the appropriate working concentration for use. Here, we screen

tetramers against CTL lines or expanded T-cell clones that are known to be specific for the MHC:peptide complex in question (this is confirmed by ^{51}CR release CTL assays or ICS assays). To determine an appropriate working concentration, make twofold serial dilutions of tetramer ranging from 1:50 to 1:1600 and use each to stain 1×10^6 antigen-specific T cells, followed by staining with anti-CD8 antibodies. Select the working concentration as the most dilute concentration that gives maximal MFI for labelled antigen-specific T cells.

4. In general, most groups report staining with tetramers at 37°C, as this appears to enhance the specificity of the tetramer–TCR interaction *(11)*. However, some variations in incubation temperature have been reported. Consequently, it may be worthwhile to initially compare tetramer staining efficiency at 4°C, room temperature and 37°C to determine optimal incubation temperatures for each individual tetramer.

5. A major benefit for using tetramers to identify antigen-specific T cells is the ability to phenotype these cells directly ex vivo. This can be achieved simply through co-staining samples with antibodies against T-cell-surface antigens, measuring the expression of these molecules on tetramer-positive T cells. Common T-cell-surface antigens that provide information on the activation state of the cell for mouse and/or human include CD69, CD25, CD44, CD62L, CD43, CD27, CD28, CD45RA/RO, CCR7, Ly6C, CD11a, $\alpha4\beta7$, CLA, P and E-selectin ligands, in addition to a growing list of chemokine receptors.

6. PI is a red fluorescent dye that binds DNA through intercalating between nucleotides. PI is membrane impermeant, and therefore excluded from live cells but not from dead cells that have lost membrane integrity. PI is therefore commonly used to detect dead cells in a population during flow cytometry.

7. The information presented here provides only a general outline for infecting stimulator cells and is likely to require virus-specific modifications for optimal viral antigen presentation. A variable parameter in this procedure is the length of time for which infected cells are incubated before use in the ICS assay. In general, viruses that have regulated gene expression (i.e. herpes viruses) require a more prolonged incubation period than those with random gene expression (i.e. influenza), as early (β) and late (γ) proteins are expressed only after hours of infection. Secondly, where stimulation with infected cells remains poor, the level of antigen presentation may be enhanced through the pre-treatment of stimulator cells with IFN-γ. This cytokine is known to enhance surface MHC class I levels *(12)* and may be particularly important when working with viruses that impair MHC pathways *(13)*.

8. Brefeldin A is a metabolite of the fungus *Eupenicillium brefeldianum* that specifically blocks protein transport from the endoplasmic reticulum (ER) to the Golgi apparatus. Consequently, BFA treatment causes T cells that are actively synthesising cytokines to retain these molecules within the intracellular compartment. The expression of intracellular cytokines is then measured by antibody staining, identifying T cells that can respond to specific antigens (i.e. antigen-specific T cells).

Acknowledgements

The authors wish to thank Professor Francis Carbone and Andrea Tarlton for their contribution of data and advice to this chapter. This work was funded by Cancer Research UK (Programme Grant C399-A2291) and the UK Medical Research Council.

References

1. Jung T, Schauer U, Heusser C, Neumann C, Rieger, C. Detection of intracellular cytokines by flow cytometry. J Immunol Methods 1993;159:197–207
2. Vikingsson A, Pederson K, Muller D. Enumeration of IFN-gamma producing lymphocytes by flow cytometry and correlation with quantitative measurement of IFN-gamma. J Immunol Methods 1994;173:219–28.
3. Murali-Krishna K, Altman JD, Suresh M, et al. Counting antigen-specific CD8 T cells: a re-evaluation of bystander activation during viral infection. Immunity 1998;8:177–87.
4. Flynn KJ, Belz GT, Altman JD, Ahmed R, Woodland DL, Doherty PC. Virus-specific CD8+ T cells in primary and secondary influenza pneumonia. Immunity 1998;8:683–91.
5. Altman JD, Moss PA, Goulder PJ, et al. Phenotypic analysis of antigen-specific T lymphocytes. Science 1996;274:94–6.
6. Klenerman P, Cerundolo V, Dunbar PR. Tracking T cells with tetramers: new tales from new tools. Nat Rev Immunol 2002;2:263–72.
7. Garboczi DN, Hung DT, Wiley DC. HLA-A2-peptide complexes: refolding and crystallization of molecules expressed in Escherichia coli and complexed with single antigenic peptides. Proc Natl Acad Sci U S A 1992; 89:3429–33.
8. Stock AT, Jones CM, Heath WR, Carbone FR. CTL response compensation for the loss of an immunodominant class I-restricted HSV-1 determinant. Immunol Cell Biol 2006;84:543–50.
9. Tscharke DC, Woo WP, Sakala IG, et al. Poxvirus CD8+ T-cell determinants and cross-reactivity in BALB/c mice. J Virol 2006;80:6318–23.
10. Hammarlund E, Lewis MW, Hansen SG, et al. Duration of antiviral immunity after smallpox vaccination. Nat Med 2003;9:1131–7.
11. Whelan JA, Dunbar PR, Price DA, et al. Specificity of CTL interactions with peptide-MHC class I tetrameric complexes is temperature dependent. J Immunol 1999;163:4342–8.
12. Schwarz K, van Den Broek M, Kostka S, et al. Overexpression of the proteasome subunits LMP2, LMP7, and MECL-1, but not PA28 alpha/beta, enhances the presentation of an immunodominant lymphocytic choriomeningitis virus T cell epitope. J Immunol 2000;165:768–78.
13. Koelle DM, Chen HB, Gavin MA, Wald A, Kwok WW, Corey L. CD8 CTL from genital herpes simplex lesions: recognition of viral tegument and immediate early proteins and lysis of infected cutaneous cells. J Immunol 2001;166:4049–58.

Chapter 2

B Cell Helper Assays

Sergio Abrignani, Elena Tonti, Giulia Casorati, and Paolo Dellabona

Abstract

Activation, proliferation and differentiation of naïve B lymphocytes into memory B cells and plasma cells requires engagement of the B cell receptor (BCR) coupled to T-cell help *(1, 2)*. T cells deliver help in cognate fashion when they are activated upon recognition of specific MHC–peptide complexes presented by B cells. T cells can also deliver help in a non-cognate or bystander fashion, when they do not find specific MHC–peptide complexes on B cells and are activated by alternative mechanisms. T-cell dependent activation of B cells can be studied in vitro by experimental models called "B cell helper assays" that are based on the co-culture of B cells with activated T cells. These assays allow to decipher the molecular bases for productive T-dependent B cell responses. We show here examples of B cell helper assays in vitro, which can be reproduced with any subset of T lymphocytes that displays the appropriate helper signals.

Key Words: Helper assay, NKT cells, B cells, Ig response.

1. Introduction

The help of CD4$^+$ T cells to B cells in secondary lymphoid organs leads to B cell expansion, the generation of short-lived antibody-secreting cells (ASCs) and, following the germinal centre reaction, to high-affinity memory B cells and long-lived plasma cells. This interaction requires presentation of processed antigens by B cells to specific CD4$^+$ T cells, signalling by costimulatory molecules such as CD40 on B cells, and production of cytokines by activated T cells. The T–B interaction can be reproduced in vitro, using an experimental system called B cell helper assay. Lanzavecchia *(1, 2)* showed that human CD4$^+$ T cell clones can activate B cells to proliferate and to differentiate into plasmacytoid ASCs that secrete Ig of all classes in vitro. B cell helper assays in vitro proved to be extremely useful to dissect the molecular mechanisms involved in the reaction. Two

Gennaro De Libero (ed.), *T Cell Protocols: Second Edition, vol. 514*

DOI 10.1007/978-1-60327-527-9_2 Springerprotocols.com

types of T-dependent help were recognised owing to different experimental designs: (i) cognate, whereby the TCR expressed by $CD4^+$ T cells interacts with cognate peptide–MHC class II complexes presented by B cells; and (ii) non-cognate, whereupon T cells, activated by third party antigen (e.g. the TCR is engaged by peptide/MHC-II expressed by other APCs, by polyclonal stimuli such as anti-CD3 mAbs, etc.), interact with B cells that do not display their cognate peptide/MHC-II complexes, yet deliver costimulatory signals sufficient to trigger B cell activation in a bystander manner. In both experimental conditions T cells are armed to induce B cell proliferation and differentiation via CD40 ligand and cytokine production. We have exploited B cell helper assays to investigate in vitro the helper function of a non-conventional subset of T lymphocyte, called invariant (i)NKT cells. Unlike conventional peptide-specific and MHC-restricted T cells, iNKT cells recognise lipid antigens presented by CD1d molecules, which are non-MHC encoded class I-like molecules *(3)*. iNKT cells express a characteristic conserved semi-invariant TCR α chain, composed by the homologue invariant Vα14-Jα18 and Vα24-Jα18 rearrangements paired with Vβ8.2, Vβ7 or Vβ2 and Vβ11 in mice and men, respectively *(4)*. Both human and mouse iNKT cells can be unequivocally identified and purified exploiting their TCR: human cells are identified using either a combination of anti-Vα24 and anti-Vβ11-specific mAb or fluorescent multimers of human recombinant soluble CD1d in complex with the glycosphingolipid αGalactosylCeramide (αGalCer) *(5, 6)*. Mouse iNKT cells are identified using fluorescent multimers of mouse recombinant CD1d molecules loaded with αGalCer *(7, 8)*. Upon activation, iNKT cells produce swiftly many different cytokines and upregulate CD40L; in turn, they induce the activation of cellular effectors of the innate (MΦ, DC, NK) as well as acquired (T, B) immune system *(4, 9)*. iNKT can be divided into two major subsets, depending on the expression of CD4 coreceptor *(10, 11)*. The iNKT $CD4^+$ subset exhibits, particularly in humans, prominent helper effector functions, characterised by the expression of a Th0 cytokine pattern, whereas the iNKT $CD4^-$ one exhibit a more polarised Th1-like effector functions and are less efficient in providing B cell help. We have utilised B cell helper assays in vitro to demonstrate that both human and mouse iNKT cells activate human or mouse B cells, respectively, to proliferate and differentiate into IgM and IgG ASCs *(12, 13)* (and E. Tonti unpublished data). We outline here the experimental procedures of the B cell helper assay, consisting essentially in the: (i) Generation of clones or polyclonal lines of human and mouse iNKT cells, respectively; (ii) Purification of primary resting human and mouse B cells; and (iii) iNKT and B cells co-culture system to investigate both cognate and bystander help provided by iNKT cells to B cells. T cell-induced B cell proliferation can be quantified by determining the incorporation of ^{3}H-thymidine into the DNA synthesised by dividing cells.

Alternatively, B cells proliferation in response to iNKT helper activity in vitro can be investigated by labelling B cells with carboxyfluorescein diacetate succinimidyl ester (CFDA-SE) and determining the progressive dilution of the fluorescent dye upon cell-division. Furthermore, B cells activated by T cells in vitro differentiate into plasmablasts that upregulate CD27 and downregulate CD20, allowing their identification and quantification by flow cytometry. This method is particularly informative when coupled with the use of CFDA-SE dilution to determine B cell division in vitro T cell helper activity in vitro leads not only to B cell division but also to differentiation to ASCs. The degree of differentiation of B cells to ACSs in vitro can be quantified by determining the concentration of secreted Igs in the culture supernatant. Finally, B cell helper assays in vitro can be performed with mouse cells as well. Mouse B cells are purified from the spleen using anti-CD19 mAb coupled magnetic MicroBeads. Mouse iNKT cells can be purified from spleen following two different methods. In the first, iNKT cells can be sorted exploiting their expression of the NK receptor NK1.1 (CD161). In this case, one has to be aware that iNKT cells represents only up to 30% of total NK1.1$^+$ T cells in the spleen. We have found that a previous depletion of NK cells in vivo, by injecting into mice the NK-specific anti-AsialoGM1 antiserum, facilitates greatly the subsequent purification in vitro of NKT cells. In the second method, iNKT cells are identified from the other NK1.1$^+$ T cells by the binding of αGalCer-loaded CD1d–IgG1 dimers to their TCR. Dimers bound to iNKT cells are revealed using a PE-labelled rat-anti-mouse IgG1 mAb and stained cells are sorted by fluorescent-based automated cell sorting or by magnetic microbeads coupled to anti-PE mAb.

2. Materials

2.1. Buffers, Media and Supplements for Cell Culture

1. RPMI complete medium: RPMI–Glutamax medium (Gibco) containing 10% FCS (HyClone), non-essential amino acids, 1 mM sodium pyruvate and kanamycin or penicillin plus streptomycin (Gibco), 2.5×10^{-5} M 2-βmercapto ethanol (2-ME).
2. FACS-wash: PBS containing 1% FCS and 0.05% NaN_3, filtered with 0.2 μm filters and kept at +4°C
3. FACS-sorting: PBS containing 1% FCS filtered with 0.2 μm filters and kept at +4°C
4. Recombinant human cytokines: IL-2 (Novartis), IL-4 and IL-13 (R&D Systems). Once lyophilised cytokines are dissolved in complete medium they can be aliquoted and stored

at –80°C over 1 year. Working aliquots can be kept at +4°C for not more than 2–3 months.

5. Recombinant mouse cytokine: IL-7 (R&D Systems) dissolved and stored as described above.
6. ^{3}H-Thymidine (^{3}H-TdR) (1 μCi/well, 6.7 Ci/mol; Amersham Corp.), keep at +4°C.
7. αGalactosylCeramide (Alexis) is dissolved at a concentration of 200 μg/ml in 0.9% (w/v) NaCl + 0.5% (v/v) polysorbate (Tween 20, Sigma Aldrich) and kept at 37°C for 16 h. To completely dissolve αGalCer, DMSO is added to the solution at a final concentration of 2% (v/v). The solution is kept at 60°C for 15 min then aliquoted in glass tubes and stored at 4°C. This αGalCer solution can be kept at 4°C for several months since its activity is very stable. Before use, each aliquot is heated at 65°C for 10 min, vortex for 3 min, and then kept at 37°C for not less than 60 min.
8. CarboxyFluorescein Diacetate Succinimidyl Ester (CFDA-SE, Molecular Probes, Leiden, The Netherlands) dissolved in DMSO at 50 mM. Stock aliquots of this solution can be prepared in polypropylene microtubes and stored at –20°C in the dark. When properly protected from the light, CFDA-SE aliquots can be thawed and frozen several times.
9. C1R LCL cells engineered with hCD1d *(14)*. Cells are grown in complete RPMI medium supplemented with 0.5 mg/ml G418. Wilde type (wt) C1R LCL cells are grown in complete RPMI medium.
10. ACK solution: 150 mM NH_4Cl, 10 mM $KHCO_4$, 0.1 mM Na_2EDTA pH 7.2–7.4 filter sterile, keep at +4°C.

2.2. Antibodies

1. Specific monoclonal antibodies used with human cells:
   a. Non-immune mouse IgG (Becton Dickinson)
   b. Anti-Vβ11 (C21, mouse IgG2a), anti-Vα24 (C15, mouse IgG1) (Immunotech, Marseille, France).
   c. Anti-CD1d (CD1d42, mouse IgG1), anti-CD3 (SK7, mouse IgG1), anti-CD3 (UCHT1, mouse IgG1), anti-CD4 (RPA-T4, mouse IgG1), anti-CD8 (SK1, IgG1), anti-CD14 (M5E2, mouse IgG2a), anti-CD20 (L27, mouse IgG1), anti-CD27 (L128, mouse IgG1), anti-CD161 (BX12, mouse IgG1) (Becton Dickinson).
   d. Agonistic anti-CD40 (626.1, mouse IgG1, Novartis Vaccine), or agonistic anti-CD40 MAB89 (mouse IgG1, BECKMAN COULTER).
   e. Anti-human CD19 MicroBeads (Miltenyi Biotec).

2. Specific monoclonal antibodies used with mouse cells:
   a. Non-immune purified rat IgG (Becton Dickinson).
   b. Anti-mouse CD3 mAb (145.2C11, hamster IgG), anti-mouse NK1.1 (PK135, mouse IgG1) (Becton Dickinson).
   c. Anti-AsialoGM1 lyophilised antiserum from Cedarline is resuspended in 1 ml of H_20 and it is stable at 4°C for up to 1 year.
   d. Mouse CD1d–Ig dimers (Dimerix Technology, Becton Dickinson).
   e. Anti-mouse CD19 MicroBeads (Miltenyi Biotec).
   f. Anti-PE MicroBeads (Miltenyi Biotec).
3. Antibodies used for human Ig ELISA:
   a. Purified anti-human IgM clone JDC-15 (capture) and biotin-conjugated clone G20-127 (detection) (Becton Dickinson).
   b. Purified polyclonal goat-anti-human IgG (capture) and biotin-conjugated goat anti-human IgG (detection) antisera (Southern Biotechnology).
   c. Purified anti-human IgG1 clone G18-145 (capture) and biotin-conjugated clone G17-1 (detection).
   d. Purified anti-human IgE clone G7-18 (capture) and biotin-conjugated clone G7-26 (detection).
   e. Total human IgM, IgG, IgG_1, (from Sigma) and IgE (Calbiochem, Darmstadt, Germany) are used as standards.
4. Antibodies used for mouse Ig ELISA:
   a. Purified polyclonal goat-anti-mouse IgM, IgG, IgG_1, $IgG2_{a/c}$ (capture) and biotin-conjugated goat anti-mouse IgM, IgG, IgG_1, $IgG2_{a/c}$ (detection) antisera (Southern Biotechnology).
   b. Total mouse IgM, IgG, IgG_1, $IgG2_{a/c}$ (from Sigma) are used as standards.

3. Methods

3.1. Purification of PBMCs

1. Heparinized peripheral blood from healthy volunteers is diluted 1:3 in RPMI–1% FCS, and layered over 2 × volume of Ficoll (Amersham).
2. The discontinuous gradients are generated by spinning at 400 *g* for 30 min at RT with the brake switched off.

3. Remove and discard the upper plasma layer without disturbing the ring of mononuclear cells at the plasma–Ficoll interface.
4. With a new pipette remove and transfer the PBMCs ring into a new tube.
5. Dilute the collected PBMCs with 4 volumes of RPMI–1% FCS and centrifuge at 180 *g* for 10 min at +4°C (brake on), to remove residual platelets.
6. Discard the supernatant, wash PBMCs once again with HBSS, and centrifuge at 200 *g* for 10 min at +4°C.
7. Discard the supernatant and resuspend the PBMC pellet in PBS–5% FCS at the appropriate concentration (see below).

3.2. Generation of Human iNKT Cell Clones and Lines

Human iNKT cells can be obtained by two methods: (i) Cell sorting and cloning ex vivo; (ii) Expansion of iNKT polyclonal cell lines in vitro with the specific antigen αGalCer. iNKT cell clones can be obtained by cloning these cell lines at limiting dilution.

3.2.1. Cell Sorting and Cloning of Human iNKT Cells *Ex Vivo*

1. Human PBMCs are incubated with PBS+300 ng/ml non-immune mouse IgG for 10 min at RT to block unspecific binding, then with saturating amounts of anti-CD3, anti-Vα24, anti-Vβ11 mAbs on ice. The amount of mAb is determined by titering each of them in separate staining.
2. PBMCs are washed three times with FACS-sorting solution and resuspended in PBS–5% FCS at 20–30 $\times$ 10^6 cells/ml. iNKT cells are identified as Vα24 and Vβ11 double positive cells within the $CD3^+$ gate. Anti-CD4 mAb can be added to the staining combination to specifically sort $CD4^+$ iNKT cells, the subset preferentially endowed with helper function.
3. iNKT cells are sorted at 1 cell/well in U-shaped 96-well plates containing 200 μl of RPMI-complete medium, 0,5 μg/ml PHA and 50 U/mL rhIL-2 and 1 $\times$ 10^6/ml pooled PBMCs obtained from 2 to 3 different donors and irradiated at 8,000 rad.
4. iNKT cell clones are grown at 37°C and 6% CO_2. Growing iNKT cell clones are phenotyped for the expression of Vα24, Vβ11 and CD4 coreceptor by flow-cytometry.

3.2.2. Expansion of iNKT Cell Polyclonal Cell Lines with the Specific Antigen αGalCer *In Vitro*

1. PBMC are plated at 10^6 cells/ml at 24-well plates in 2 ml of complete medium containing 10–100 ng/ml of αGalCer. After 48 h, 20 U/ml of rhIL-2 (Novartis, Emeryville, CA, US) are added. IL-2 is then added again at 40 U/ml every 48 h (*see* **Note 1**).
2. At day 15–20, αGalCer-induced cell lines are stained with anti-Vβ11, Vα24, and CD4 mAbs as described above (*see* **Note 4**);
3. iNKT cells are sorted by high-speed cell sorting either as bulk cell line or as clones. $V\alpha24^+V\beta11^+CD4^+$, $V\alpha24^+V\beta11^+CD4^-$ T

cells are sorted/cloned at 1 cell/well in 96-well plates containing 200 μl of RPMI-complete medium, 0.5 μg/ml PHA and 50 U/ml rhIL-2 and 1×10^6/ml pooled PBMCs obtained from 2 to 3 different donors and irradiated at 8,000 rad.

3.2.3. Propagation of Human iNKT Cell Clones *In Vitro*

1. iNKT clones or lines are propagated by restimulation every 15–20 days in RPMI-complete medium with a mixture of irradiated allogeneic feeder cells (see above) at 1×10^6 cells/ml, PHA (0.5 μg/ml) and 50 U/ml rhIL-2.

3.2.4. Antigen-Specificity Assays for iNKT Cells

1. To verify the antigen-specificity of the iNKT cell clones, 2.5×10^4 cells are plated in U-bottom 96-well plates with 5×10^4 of either C1R-CD1d or C1R wt cells in the presence of 50 ng/ml of αGalCer.
2. After 48 h, antigen-specific iNKT cell activation is determined by measuring the concentration of IFN-γ and IL-4 in the culture supernatant by ELISA.

3.3. Human B Lymphocyte Purification

1. Human PBMCs are obtained from heparinized peripheral blood of healthy volunteers after density gradient centrifugation on Ficoll–Paque (see above, *see* **Notes 2 and 3**).
2. For B cell purification, PBMC are first incubated with PBS+10 μg/ml non-immune mouse IgG for 10 min at RT to block unspecific binding, then anti-CD19 mAb conjugated with magnetic beads are added for further 20 min at 4°C according to manufacture instruction (Miltenyi). Immunomagnetically labelled B cells are purified on LS+ columns using the MACS® separation system (Miltenyi). All B lymphocyte preparations should be >95% $CD20^+$, the few contaminating cells being $CD14^+$.

3.4. Human B Cell Helper Assays

3.4.1. B Cell Proliferation Determined by ^{3}H-Thymidine Incorporation

1. iNKT cell clones are irradiated at 6,000 rad.
2. Purified B lymphocytes ($0.5–1 \times 10^5$ cells) are plated alone or with irradiated cells of iNKT cell clones at 2:1 ratio in U-bottom 96-well plates containing 0.2 ml of RPMI-complete medium. Duplicate or triplicate wells are stimulated with either 0.2 μg/ml anti-CD3 mAb (UCHT-1) or 50 ng/ml of αGalCer or equal amounts of mouse IgG1 or vehicle.
3. After 5 days in culture, 0.5–1 μCi of ^{3}H-thymidine is added to each well for 16 h, before cell harvesting and scintillation counting.

3.4.2. B Cell Proliferation Determined CFDA-SE Dilution Assay

1. Purified B cells are loaded with CFDA-SE to give a final concentration ranging between 0.5 and 5 μM, depending on the duration of the study of cell tracking. Generally, longer

duration studies required more intense staining to abrogate the effects of fluorescence decay due to catabolic processes occurring in the absence of cell division *(15)*. B cells are incubated for 8 min at room temperature at a cell concentration of 2×10^7/ml (*see* **Note 4**).

2. Labelling is stopped by adding an equal volume of FCS, and washing the cells three times in complete medium.
3. CFDA-SE labelled B lymphocytes are seeded at 5×10^5/ml in U-bottom wells containing complete medium, with or without 10 µg/ml of agonistic anti-CD40 (626.1, or MAB89) or 50 ng/ml αGalCer, in the presence or in the absence of non-irradiated iNKT cells, at a ratio of 20 B cells to 1 NKT cell.
4. After 5 days, CFSE dilution in B cells is determined by four colour-cytofluorimetric analysis using staining for: CFDA-SE, CD27, CD20 and CD3 (to gate out T cells).

3.4.3. Antibody Production *In Vitro*

1. Primary autologous or allogeneic B cells and irradiated iNKT cells are cocultured at 2:1 ratio as described above.
2. On day 10, culture supernatants are collected and used in ELISA to determine immunoglobulin concentration. IgM, IgG and IgE in the culture supernatants are measured with a sandwich LISA. Briefly, polystyrene microplates are coated with anti-human IgM or IgG or IgE antibodies (10 µg/ml) in coating buffer (Na_2CO_3 0.18 g/l, $NaHCO_3$ 2.1 g/l, NaCl 9 g/l, NaN_3 0.2 g/l), pH 8.6 at 4°C overnight. The plates are washed and serial dilutions of the culture supernatants in PBS–FCS are added for 2 h at room temperature. After washing, anti-human IgM or IgG or IgE antibody, coupled to biotin, is added for a further 2 h. After washing, AP-conjugated Strepatividin (Becton Dickinson) is added to wells for 30 min at RT. After final washing, the enzyme bound to the wells is determined using 1 mg/ml *p*-nitrophenyl phosphate (Sigma Chemicals, St. Louis, MO) in 1 M ethanol amine buffer, pH 9.6. The absorbance is read with an automated photometer. The amount of Ig is determined by comparison with a standard curve constructed using known amounts of purified IgM, IgG and IgE.

3.5. Mouse B Cell Helper Assays with iNKT Cells

3.5.1. Mouse B Cell Purification

1. Spleens are obtained from mice under sterile conditions and minced into small fragments by scalpel.
2. Fragments are gently homogenised on sterile 70 µm cell strainers (Falcon) to obtain a single cell suspension.
3. Cells are centrifuged 1 × at 200 *g* for 5 min at +4°C.
4. Resuspend pellet in 10 ml PBS+2% FCS. Add 3 ml ACK solution for 2 min at RT to lyse RBCs.

5. Wash 2 × and resuspend the pellet at 5×10^6 cells/ml in PBS–1% FCS and incubated with immunomagnetic anti-CD19 magnetic MicroBeads according to manufacturer's instructions. Usually, one spleen yields $>5 \times 10^7$purified B cells.

3.5.2. Mouse iNKT Cell Purification

1. To deplete NK cells from mice, 40 μl of anti-asialoGM1 anti-serum is injected i.p. 2 days before iNKT cell purification (*see* **Note 4**).
2. Spleen is taken from mice and processed to obtain a single cell suspension as described above.
3. Spleen cells are deprived of adherent cells by 1 h culture on bacteriological petri dishes in RPMI–1% FCS at 37°C.

3.5.2.1. Mouse NKT Cell Purification Using Anti-NK1.1 mAb

1. To sort total NKT cells, the non-adherent fraction of spleen cells is stained with a saturating dose (pre-determined in pilot experiments) of PE-conjugated anti-NK1.1 mAb for 20 min at 4°C.
2. Cells are washed 3 × with PBS–1% FCS.
3. NK1.1-PE-labelled total NKT cells incubated with anti-PE magnetic MicroBeads according to manufacturer's instructions.

3.5.2.2. Mouse Primary iNKT Cell Purification Using αGalCer-Loaded CD1d-Dimers

1. mCD1d-DimeriX (Becton Dickinson) are loaded with aGal-Cer o.n. at RT (*see* **Note 5**). Usually, 1 μg dimers/staining is sufficient (*see* **Note 6**).
2. Non-specific binding sites on cells are blocked by incubation with 300 ng of purified non-immune Rat IgG for 20 min at RT.
3. To stain iNKT cells, the non-adherent fraction of spleen cells is stained for 60 min on ice of αGalCer-loaded CD1d-dimers in FACS-sort.
4. Wash three times with FACS sort.
5. After washing, αGalCer-loaded CD1d-dimers are revealed by incubations with anti-mouse IgG1-PE mAb (A85-1) for 10 min at +4°C.
6. Cells are washed and non-specific antibody binding sites are blocked by incubating 20 min at +4°C with 300 ng each of mouse and rat IgG1 molecules.
7. Without washing, cells are incubated with anti-TCR-β-FITC (H57-597) for 10 min at +4°C.
8. iNKT cells are purified using standard fluorescent-based automated cell sorting.

3.5.2.3. Mouse iNKT Cell Purification Using αGalCer-Loaded CD1d-Dimers After Expansion In Vitro

1. Total spleen cells are plated in wells of 24 well plates at 10^6 cell/ml in 2 ml of compete medium containing 50 ng/ml of αGalCer.
2. After 24 h add 50 U/ml of rhIL-2 and 20 ng/ml rmIL-7.
3. Replace fresh cytokines every 48 h.
4. At day 12 of culture, stain growing iNKT cells with αGalCer-loaded CD1d-dimers and proceed to purification using standard fluorescent based automated cell sorting as described above.

3.5.3. Mouse B Cell Helper Assays with Purified Total Splenic NKT Cells (Purified with Anti-NK1.1 Immunomagnetic Sorting)

1. One day before the B cell helper assay, insolubilise the purified anti-CD3 mAb 1452C11 on wells of round-bottom 96-well plates. To each well add 25 μl of a 5 μg/ml solution of the mAb in PBS, and incubate at +4°C. Alternatively, the anti-CD3 mAb can be insolubilised the same day of the B cell helper assay by incubating 2 h at 37°C.
2. Wash wells 4 × times with PBS–5% FCS.
3. Irradiate the purified $NK1.1^+$ T cells at 3,000 rad.
4. Plate the purified B cells (5×10^5) with total irradiated $NK1.1^+$ T cells (5×10^4) in wells pre-coated with the anti-mouse CD3 mAb in 200 μl of complete medium.
5. On day 10, collect culture supernatant and determine the concentration of secreted Igs by ELISA.

3.5.4. Mouse B Cell Helper Assays with Purified Splenic iNKT Cells (Purified by Using αGalCer-Loaded CD1d-Dimers and Cell Sorting)

1. One day before the B cell helper assay, insolubilise purified goat anti-rat IgG (mouse and human adsorbed) on wells of round-bottom 96 well plates. This antiserum will cross-link the rat-anti mouse IgG1-Pe mAb utilised to reveal the αGalCer-CD1d-dimers bound to iNKT cells. iNKT cells will be activated upon cross-linking the CD1d-dimers bound to the invariant TCR. To each well add 25 μl of a 10 μg/ml solution of the goat anti-rat IgG Ab in PBS, and incubate at +4°C. Alternatively, the goat anti-rat Ab can be insolubilised the same day of the B cell helper assay by incubating for 2 h at 37°C.
2. Wash wells 4 × times with PBS–5% FCS;
3. Irradiate the purified iNKT cells at 3,000 rad;
4. Plate the purified B cells (5×10^5) with irradiated iNKT cells (5×10^4) in wells pre-coated with the goat anti-rat Ab in 200 μl of complete medium;
5. On day 10, collect culture supernatant and determine the concentration of secreted Igs by the ASCs that have differentiated in vitro by ELISA.

6. As control for the B cell helper assays, purified B cells (5×10^5) are in 200 μl of complete medium supplemented with or without +10 μg/ml LPS.
7. On day 10, collect supernatant from the culture cultures and determine by ELISA the concentration of secreted.

3.5.5. ELISA for Mouse Ig

1. To determine the amount of IgM and IgG produced by mouse B cells in the helper assays, we follow the ELISA method identical to the one used for human Ig. The following pairs of antibody (Souther Biotechnology) specific for mouse Ig are utilised: purified goat-anti mouse IgM (capture) and biotin-conjugated goat-anti-mouse IgM (detection); purified goat-anti mouse IgG (capture) and biotin-conjugated goat-anti-mouse IgG (detection). The insoluble immune complexes are revealed with AP-conjugated streptavidin as described above for the ELISA detection of human Ig.

4. Notes

1. Human iNKT cell expansion upon αGalCer stimulation in vitro can be variable from donor to donor both in terms of total cell yield and of time required for maximal growth. It is suggested to activate two or three independent PBMCs a time, to select for the best iNKT cells expansion. On average, a time between 15 and 20 days can be considered sufficient to obtain significant iNKT cell expansion in vitro.
2. Sometimes, B cells obtained from buffy coats are poor responders or are already activated. For this reason, B cells purified from fresh whole blood of healthy donor should be preferred.
3. The use of human or mouse sera in helper assays involving human and mouse B cells, respectively, should be avoided as their presence can lead to high background values when culture supernatants are tested in ELISA.
4. CFDA-SE labelling of cells, B cells included, can be quite toxic and reduce the survival of labelled cells in vitro. It is therefore critical to titer carefully the minimal concentration of the dye necessary to track the target cells over the desired culturing time vis a vis their maximal survival.
5. 40 μl of anti-Asialo GM1 antiserum injected i.p. into mice is usually sufficient to deplete >95% of $NK1.1^+$ cells in 24 h. NK cells remain depleted for 3 days.
6. CD1d-DimerX (Becton Dickinson) are loaded with αGalCer on at RT. Efficient Dimer loading requires a 1:9 molar ratio

dimers:αGalCer. To determine the amount of αGalCer to use for each loading, apply the following formula:

$$\mu g\ di\ \alpha GalCer = \frac{\mu g\ di\ CD1d - Dimer \times 9\ (molar\ eccess\ of \alpha GalCer) \times 858\ (MW\ \alpha GalCer)}{250,000 (MW\ CD1d - Dimer)}$$

The αGalCer-Dimer solution is spun down quickly before using it for staining.

References

1. Lanzavecchia A. One out of five peripheral blood B lymphocytes is activated to high-rate Ig production by human alloreactive T cell clones. Eur J Immunol 1983;13:820–4.
2. Lanzavecchia A, Parodi B, Celada F. Activation of human B lymphocytes: frequency of antigen-specific B cells triggered by alloreactive or by antigen-specific T cell clones. Eur J Immunol 1983;13:733–8.
3. Bendelac A, Lantz O, Quimby ME, Yewdell JW, Bennink JR, Brutkiewicz RR. CD1 recognition by mouse NK1+ T lymphocytes. Science 1995;268:863–5.
4. Bendelac A, Savage PB, Teyton L. The Biology of NKT Cells. Annu Rev Immunol 2007;25:297–336.
5. Dellabona P, Padovan E, Casorati G, Brockhaus M, Lanzavecchia A. An invariant V alpha 24-J alpha Q/V beta 11 T cell receptor is expressed in all individuals by clonally expanded CD4-8- T cells. J Exp Med 1994;180:1171–6.
6. Karadimitris A, Gadola S, Altamirano M, et al. Human CD1d-glycolipid tetramers generated by in vitro oxidative refolding chromatography. Proc Natl Acad Sci U S A 2001;98:3294–8.
7. Matsuda JL, Naidenko OV, Gapin L, et al. Tracking the response of natural killer T cells to a glycolipid antigen using CD1d tetramers. J Exp Med 2000;192:741–54.
8. Benlagha K, Weiss A, Beavis A, Teyton L, Bendelac A. In vivo identification of glycolipid antigen-specific T cells using fluorescent CD1d tetramers. J Exp Med 2000;191:1895–903.
9. Kronenberg M. Toward an understanding of NKT cell biology: progress and paradoxes. Annu Rev Immunol 2005;23:877–900.
10. Gumperz JE, Miyake S, Yamamura T, Brenner MB. Functionally distinct subsets of CD1d-restricted natural killer T cells revealed by CD1d tetramer staining. J Exp Med 2002;195:625–36.
11. Lee PT, Benlagha K, Teyton L, Bendelac A. Distinct functional lineages of human V(alpha)24 natural killer T cells. J Exp Med 2002;195:637–41.
12. Galli G, Nuti S, Tavarini S, et al. CD1d-restricted help to B cells by human invariant natural killer T lymphocytes. J Exp Med 2003;197:1051–7.
13. Galli G, Nuti S, Tavarini S, et al. Innate immune responses support adaptive immunity: NKT cells induce B cell activation. Vaccine 2003;21 Suppl 2:S48–54.
14. Brossay L, Chioda M, Burdin N, et al. CD1d-mediated recognition of an alpha-galactosylceramide by natural killer T cells is highly conserved through mammalian evolution. J Exp Med 1998;188:1521–8.
15. Lyons AB, Parish CR. Determination of lymphocyte division by flow cytometry. J Immunol Methods 1994;171:131–7.

Chapter 3

transkingdom RNA Interference (*tk*RNAi): A Novel Method to Induce Therapeutic Gene Silencing

Thu A. Nguyen and Johannes H. Fruehauf

Abstract

RNA interference is a phenomenon in which specific, endogenous genes are silenced by mRNA degradation. This technology is highly regarded as a potential therapeutic due to its high efficacy and low toxicity. However, the difficulty of delivering RNAi to target cells has impeded the development of RNAi-based therapies. One method to overcome this barrier is the use of a nonpathogenic bacteria vector, *Escherichia coli*, to deliver RNAi to target cells with high efficacy. In *transkingdom* interference RNAi (*tk*RNAi) delivery, *E. coli* were engineered to transcribe short RNA (shRNA) from a plasmid (TRIP) containing the invasin gene *Inv* and the listeriolysin O gene *Hly*. *tk*RNAi is successful in eliciting efficient gene silencing in vitro and in vivo.

Key words: RNAi, *transkingdom* RNA interference, *tk*RNAi, shRNA, siRNA, Gene therapy.

1. Introduction

RNA interference (RNAi) is a naturally occurring pathway in which fragments of double-stranded ribonucleic acid (dsRNA) or short interference RNA (siRNA) selectively downregulate genes with complementary sequences to the siRNA *(1)*. The ability of RNAi to silence genes by degrading the mRNA and preventing protein expression has many advantageous implications for the field of biology. In addition to serving as a valuable genomic tool for the assessment of gene functions, RNAi can also be used as a means of genetic therapy to combat various diseases such as viral infections, cancer, and neurodegenerative disorders. While this innovative technology holds many promises for the future, the expansion of RNAi-based therapies is hindered by

Gennaro De Libero (ed.), *T Cell Protocols: Second Edition, vol. 514*

DOI 10.1007/978-1-60327-527-9_3 Springerprotocols.com

the challenge of delivering RNAi to target cells. Although the use of viral vectors, selective particles, liposomes, and chemical modifications to siRNA has been explored in the delivery of RNAi, little is known about the use of bacteria to induce gene silencing. Yet, as a versatile gene vector, *transkingdom* RNAi has recently been shown to be an effective, safe, and inexpensive measure for delivering RNAi to target cells with high selectivity and high specificity *(2–5)*.

The new bacteria-mediated RNAi delivery system termed *transkingdom* RNA interference (*tk*RNAi) combines production and delivery of short hairpin RNA in genetically modified bacteria, e.g., *E. coli*. Through the presence of the *transkingdom* RNA interference plasmid (TRIP) (**Fig.** 3.1), *E. coli* are enabled to produce specific shRNA, invade target cells, and release shRNA into the target cell cytoplasm, where they will trigger RNA interference. The TRIP plasmid contains three key components, the shRNA expression cassette, an *invasin* gene and a *listerolysin* gene, in addition to its bacterial origin and a selection marker.

The shRNA expression cassette allows for a high-level production of shRNA under the control of a T7 RNA polymerase promoter, and is flanked by two unique restriction sites, which give convenient access to change the shRNA expression insert. This enables the user to rapidly make numerous constructs targeting different genes of interest.

The introduction of the *inv* gene from *Yersinia pseudotuberculosis* leads to the expression of invasin protein on the bacterial

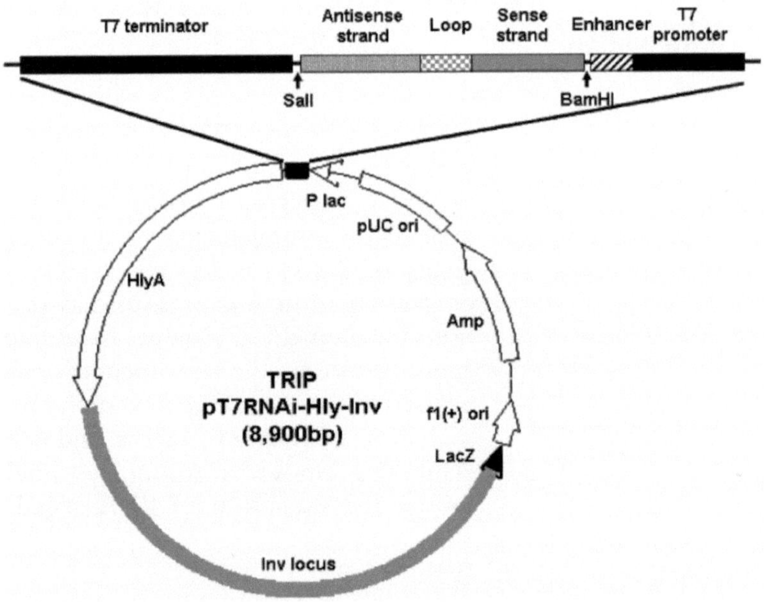

Fig. 3.1. Structure of *transkingdom* RNA interference plasmid (TRIP).

surface and gives the *E. coli* the ability to invade beta(1)integrin expressing mammalian cells with high efficiency *(4)*. TRIP also encodes for the production of listeriolysin O, a pore-forming toxin (encoded by the *hly* gene from *Listeria monocytogenes)* in the bacterial cytoplasm, which leads to the destruction of the endosome membrane after bacterial uptake into the host cell *(5–7)*. After mammalian cell invasion, bacteria will reside in an endo/lysosome and lyse either due to their lack of intracellular survival mechanisms, the addition of antibiotics, or the action of lysosomes. At this time, listeriolysin as well as shRNA will be released, and listeriolysin will destroy the endosome membrane, allowing the shRNA to escape into the host cell cytoplasm, where it becomes processed into siRNA, a mediator of the RNAi pathway.

To illustrate the effects of this technology, a bacterial plasmid termed TRIP (*transkingdom* RNAi plasmid) was constructed against the human cancer gene *CTNNB1* (catenin β-1) and introduced into BL21DE3, a competent strain of nonpathogenic *E. coli*. The transformed strain of BL21DE3 was used to treat SW480 cells (human colon cancer) in vitro. *transkingdom* RNAi is also effective in inducing local and systemic gene silencing after application in live animals. *E. coli* expressing shRNA against mouse *Ctnnb1* was administered orally to induce local gene silencing in the intestinal epithelium of mice. After intravenous administration of *E. coli* encoding shRNA against human *CTNNB1* in mice with xenografts of human colon cancer, catenin levels were decreased in the tumors on mRNA and protein levels, and tumor growth was inhibited *(2)*.

2. Materials

2.1. Trip Plasmid Construction

1. *E. coli* strain BL21DE3.
2. Oligonucleotides containing multiple cloning sites, T7 promoter, enhancer and terminator (Qiagen).
3. *CTNNB1* hairpin oligonucleotides.
4. *Inv* gene and *Hly* gene sequences.
5. Pfx DNA polymerase (Invitrogen).

2.2. Transformation of BL21DE3

1. Competent BL21DE3 from Gene Therapy Systems.

2.3. In Vitro *Transfection of SW480 Cells*

1. Brain–Heart-Infusion broth (Remel).
2. Ampicillin.
3. IPTG.

4. SW480 cells (human colon cancer cells).
5. RPMI1640 Medium (Sigma Aldrich).
6. Fetal Bovine Serum (Sigma Aldrich).
7. Antibiotics – Penicillin G, Streptomycin, and Amphotericin (Sigma Aldrich).
8. Ofloxacin (Sigma).
9. Cell culture dishes (6 cm).

2.4. In Vivo Treatment

1. Female C57/BL6 mice (Charles River Laboratories).
2. Female BALB/c nude mice (Charles River Laboratories).
3. Phosphate buffered saline (PBS)

2.5. Gene Silencing Analysis

2.5.1. RTq-PCR

1. Taqman RTq-PCR primers for human *CTNNB1*
   1 hCTNNB1 reverse: 5' – GGACAAAGGGCAAGATTT – 3'.
   2 hCTNNB1 forward: 5' – AGCTCTTACACCCACCATC – 3'
   3 hCTNNB1 probe: 5' – FAM-CTGGCCTCTGATAAAGGC – 3'.
2. Taqman RTq-PCR primers for mouse Ctnnb1
   1 mbCAT reverse: 5' – FAM –TCACGCAAGAGCAAGTAG – 3'.
   2 mbCAT forward: 5' – GAGCCCTAGTCATTGCATA – 3'.
   3 mbCAT probe: 5' – FAM-TCACGCAAGAGCAAGTAG – 3'.

2.5.2. Immunohistochemistry

1. Primary antibodies: Rabbit polyclonal β-catenin (1:250) (Santa Cruz).
2. ABC (Vector) staining kit.

2.5.3. Northern Blot Analysis

1. Antisense strand probe: 5' – GUAGCUGAUAUUGAUGGA-CAG – 3'.

2.5.4. Western Blot Analysis

1. Primary antibodies: Rabbit polyclonal B-catenin (1:500) (Santa Cruz).
2. Secondary antibodies: Goat anti-rabbit (1:100) (Santa Cruz).

3. Methods

The following methods describe (1) the construction of the TRIP plasmid, (2) transformation of TRIP into *E. coli* bacteria, (3) in vitro treatment of SW480 cells, and (4) in vivo experiments involving oral treatment to induce gene silencing in intestinal epithelial cells and systemic dosing with intravenous treatment.

3.1. TRIP Plasmid Construction

1. Oligonucleotides containing multiple cloning site (MCS), T7 promoter, enhancer, and terminator (synthesized by Qiagen) were ligated into blunted *Bss*HII sites of KSII(+).
2. shRNA encoding sequence (*CTNNB1* hairpin oligonucleotides) was inserted into the *Bam*HI and *SAlI* sites of the KSII(+) produce plasmid pT7RNAi against *CTNNB1*.
3. *hly* gene was amplified from pGBΩ2*inv-hly* (kindly provided by C. Grillot-Courvalin) by PCR (Pfx DNA polymerase from Invitrogen) and with the use of hly-1 primer (5'-CCCTCCTTT-GATT AGTATATTCCTATCTTA-3') and hly-2 primer (5'-AAGCTTTTAAATCAGCAGGGGTCTTTTTGG-3'), and cloned into the *Eco*RV site of KSII(+) to make KSII(+)/*hly*.
4. The *Pst*I fragment containing the *inv* locus of pGBΩ2*inv-hly* was inserted into the *Pst*I site of KSII(+)/*hly*.
5. The *hly-Inv* fragment was excised with *Bam*HI and *Sal*I and blunted.
6. The blunted *Hly-Inv* fragment was ligated into the EcoRV site in the T7 terminator of the pT7RNAi to make TRIP (*see* **Note 1**).

3.2. Transformation of BL21DE3

1. Transform TRIP plasmid into BL21DE3 strain using standard heat shock protocol.

3.3. In Vitro *tkRNAi* Treatment of SW480 Cells

3.3.1. Bacterial Culture

1. Plate cells on agar plates containing 100 μg/ml ampicillin and incubate overnight at 37°C.
2. Select one colony and grow overnight in Brain–Heart-Infusion-broth (Remel) with 100 μg/ml ampicillin at 37°C, the night before cell transfection.
3. Inoculate overnight cultures of bacteria into fresh medium and grow for 2 h for effective transfection..
4. Add IPTG (isopropyl-B-D-thio-galactopyranoside), an activator of T7 RNA polymerase, and incubate for an additional 1–2 h.
5. Measure bacteria count at OD_{600} with a spectrophotometer (*see* **Note 2**).

3.3.2. Cell Culture and Treatment Protocol for tkRNAi

1. Culture SW480 cells into 6-cm plates at 30% confluence 18 h prior to treatment with bacteria.
2. Culture SW480 cells in RPMI1640 medium supplemented with antibiotics (100 U/ml penicillin G, 10 μg/ml streptomycin, and 2.5 μg/ml amphotericin) and 10% fetal bovine serum or FBS (Sigma).
3. Thirty minutes before the addition of bacteria for cell transfection, remove the old medium supplemented with FBS and antibiotics and replace with fresh medium without serum or antibiotics.
4. Meanwhile, spin down the bacteria while in early log phase at approximately 3,000 rpm for 10–15 min or until the bacteria has

compacted into a pellet. Remove BHI medium, wash the bacteria in PBS, dilute in RPMI medium without serum and antibiotics and add to cells at the desired multiplicities of infection or MOIs.

5. Incubate cells with *E. coli* at varying MOIs for 1–2 h.
6. Wash cells extensively and treat with antibiotics to remove intracellular and extracellular bacteria (*see* **Note 3**).
7. Culture cells for an additional 24–96 h before harvesting.
8. Analyze gene silencing.
9. Include appropriate controls (untreated and vector treated).

3.4. In Vivo *Treatments with tkRNAi*

3.4.1. Oral Treatment for Local Silencing

1. Randomize female C57/BL6 mice into required groups, include appropriate controls (vehicle control, PBS control)
2. Orally administer the treatment group with *E. coli* expressing shRNA against mouse *Ctnnb1* re-diluted in PBS.
3. Orally administer the vehicle control group with *E. coli* containing the TRIP vector against a control gene (e.g., GFP).
4. Treat both groups daily (up to 5 days per week) for the desired period (*see* **Notes 4**).

3.4.2. Treatment of Xenograft Tumors Through iv Treatment with tkRNAi

1. Subcutaneous implantation of SW480 cells into the flank regions of female BALB/c nude mice approx 10 days prior to treatment. Randomize mice into appropriate treatment groups. Include appropriate controls (at least vector control, untreated control)
2. Intravenous injections (tail vein injection) of the treatment group with 10^6 cfu *E. coli* expressing shRNA against *CTNNB1* re-diluted in 200 μl PBS
3. Intravenous injections of the vector control group with *E. coli* containing the TRIP vector against a control gene (e.g., GFP).
4. Treat both groups three times per week for 3 weeks.

3.5. Analysis

1. RTqPCR to quantify and determine changes in the levels of target gene RNA.
2. Western blot to determine target gene (and downstream) protein concentration.
3. Northern blot for detection and analysis of shRNA.
4. Immunohistochemistry staining to determine the distribution and localization of specific biomarkers.

4. Notes

1. Activation of the *CTNNB1* cancer gene by means of overexpression or random mutations can lead to the development of a number of cancers, especially colon cancer. This pathway is

widely targeted in experiments pertaining to the development of novel RNAi techniques since this cancer gene is often stabilized in cells. In terms of the TRIP plasmid, the production of shRNA is controlled by the bacteriophage T7 promoter and not by the promoter or enhancer of host mammalian cells. Furthermore, T7 RNA polymerase is provided by BL21DE3 to transcribe shRNA from TRIP. The *Inv* locus of the TRIP vector plays a key role in the admission of noninvasive *E. coli* into the β-1 integrin positive mammalian cells while the product of the *Hly* gene (listeriolysin O) initiates the release of genetic materials from vesicles *(1)*.

2. It is recommended that one colony of bacteria is grown overnight in a 5 mL volume of BHI medium with ampicillin at 50–100 μg/ml. For optimal bacteria growth, 1% of the overnight culture should be inoculated, under sterile conditions, into a desired volume of BHI medium containing ampicillin and shaken for approximately 2 h at 37°C. After 2 h, add IPTG to the bacteria culture to maximize the transcription rate and incubate at 37°C for an additional 1–2 h. Best efficacies were observed with bacteria measured in the OD_{600} range of 0.8–1.1 than at other ODs.
3. For optimal transfection with *E. coli* while ensuring maximum cell survival rate, cells should be washed approximately 2–3 times with PBS and 2–3 times with complete medium (RPMI1640 medium containing serum and antibiotics) following the bacterial infection. Gene silencing potency increased with incubation times up to 2 h. Following the washes, add fresh, complete medium and Ofloxacin at 10 μg/ml. The potency of gene silencing is dependent on MOI, with near-complete gene silencing observed at an MOI of 1:1,000.
4. A dosage of approximately 10^9–10^{10} colony-forming units (c.f.u.) is usually well tolerated for oral application. Treatment schedules can be optimized by monitoring bacterial shedding in the stools. Good treatment effects have been observed with the number of TRIP-carrying bacteria in the stools in the order of 5% and greater (this may allow to reduce the frequency of treatment).

References

1. Fire A, Xu S, Montgomery MK, Kostas SA, Driver SE, Mello CC. Potent specific genetic interference by double-stranded RNA in *Caenorhabditis elegans*. Nature 1998;391: 806–11
2. Xiang S, Fruehauf JH, Li CJ. Short hairpin RNA-expressing bacteria elicit RNA interference in mammals. Nature Biotechnology 2006;4: 697–702.
3. Li CX, Parker A, Menocal E, Xiang S, Borodyansky L, Fruehauf JH. Delivery of RNA interference. Cell Cycle 2006;5:2103–9.
4. Isberg RR, Leong JM. Multiple beta 1 chain integrins are receptors for invasin, a protein

that promotes bacterial penetration into mammalian cells. Cell 1990;60:861–71
5. Palffy R, Gardlik R, Hodosy J, et al. Bacteria in gene therapy: bactofection versus alternative gene therapy. Gene Therapy 2005;10:1–5.
6. Grillot-Courvalin C, Groussard S, Huetz F, Ojcius DM, Courvalin P. Functional gene transfer from intracellular bacteria to mammalian cells. Nature Biotechnology 1998;16: 862–6.
7. Wing EJ, Gregory SH. *Listeria monocytogenes*: Clinical and experimental update. Journal of Infectious Diseases 2002;185: 18–24.

Chapter 4

Flow Cytometry and Cell Activation

Sonia Gavasso

Abstract

Flow cytometry is combined with highly specific fluorophore-conjugated antibodies that will only bind to the activated forms of molecules. The advances in flow cytometry enable to perform quantitative multiplexed analysis of single cells within heterogeneous populations stained with specific antibodies for phenotyping in conjunction with antibodies to phosphorylated, i.e., activated molecules within signaling pathways. By reactivating signaling pathways in vitro it is possible to collect data on the responsive state of complex cell populations such as immune cells. In this protocol, peripheral blood mononuclear cells (PBMC) are stimulated with cytokines for the indicated time in a 37°C/CO_2 incubator, fixed immediately with paraformaldehyde to freeze signaling, permeabilized with methanol, and then stained simultaneously with an antibody cocktail to signaling molecules within the JAK-STAT pathway and phenotypic markers for T-cells and B-cells. The protocol shows a basic four-color method which can be expanded to potentially study any signaling pathway in a defined cell subset.

Key words: Flow cytometry, PBMC, phosphoprotein, immune cells, signaling, phospho-antibodies, activation, multiparameter, phosphorylation.

1. Introduction

Flow cytometry is developing into an exciting tool for the study of post-translational modifications at the single cell level. The advances in hardware and software, in parallel with commercially available reagents enable analysis of signaling networks in primary samples such as peripheral blood mononuclear cells (PBMC). Highly specific fluorophore-conjugated antibodies are combined with flow cytometry to perform quantitative multiparameter analysis of single cells within complex cell population *(1)*.

Gennaro De Libero (ed.), *T Cell Protocols: Second Edition, vol. 514*

DOI 10.1007/978-1-60327-527-9_4 Springerprotocols.com

The potential to combine flow cytometry with fluorophore-conjugated antibodies to specific activated molecules within a particular signaling cascade makes this technology an innovative tool to analyze the responsive state of immune cells. Immune cells communicate through signaling molecules such as cytokines in order to coordinate a response to pathogens. Intracellularly, these signals are among others propagated by phosphatases and kinases to induce an appropriated cell type specific response *(2, 3)*. By combining antibodies that only recognize the phosphorylated forms of molecules and traditionally used markers for immunophenotyping with flow cytometry new approaches to phosphoprotein analysis have become available. The technique enables analysis of intricate signaling networks within complex cell populations without the need to separate the cells of interest first. It is possible to gather data on the pertinent signaling pathway in the cell population of interest by reactivating particular pathways in vivo or in vitro. Flow cytometry has the unique capability to produce quantitative data of these phosphorylation events at the single cell level.

For this technique to work successfully, experience in flow cytometry is recommended. Besides instrument configuration in your particular cytometer it is also crucial to be familiar with excitation and emission of particular fluorophores as well as their chemical properties, i.e., stability. Furthermore, compensation is inevitable when running a multi-color experiment *(4)*. Many fluorophore-antibody conjugates are being developed and tested specifically for flow cytometry and often the antibodies are commercially available, already conjugated to fluorophores.

To illustrate the method PBMC are being stimulated with a panel of cytokines, fixed, permeabilized, and stained with cell type specific markers for T-cells (CD3+) and B-cells (CD20+) in conjunction with antibodies to phosphorylated intracellular signaling molecules of the JAK-STAT signaling cascade central to immune function (Janus Kinases, Signal Transducer and Activator of Transcription). The protocol is an example of a multiplexed four-color analyses within a heterogeneous cell population. For further reading see *(5–11)*.

2. Materials

1. Separation of mononuclear cells from whole blood. For example BD Vacutainer CPT tubes (*see* **Note 1**)
2. PBS, cell culture grade

3. Serum-free media that supports PBMC such as BioWhittaker X-Vivo 15
4. Falcon tubes (15 ml)
5. Hemocytometer
6. Trypan blue
7. Cryovials (4°C)
8. DMSO, cell culture grade
9. Serum-free cryopreservation media such as ProFreeze-CDM serum-free media, 4°C
10. Freezing container for repeatable cooling rate such as "Mr. Frosty" Freezing container, Nalgene
11. Isopropanol
12. Warm paraformaldehyde (37°C), 16% paraformaldehyde ampoules, Electron Microscopy services, store at room temperature (RT), keep away from light sources and use within 1 week. Caution: Paraformaldehyde is toxic by inhalation and ingestion. Handle with care and use appropriate protective measurements (*see* **Note 2**)
13. Methanol, 95%. Store at –20°C.
14. Cell culture tested stimuli such as cytokines : INF-β, INF-γ, IL-2, IL-10, IL-6 (*see* **Note 3**)
15. Staining buffer: PBS with 1% BSA, pH 7.4, filtered 0.2 μm. It is recommended to make fresh staining buffer (*see* **Note 2**)
16. Prepare cytokine or stimuli working solutions (*see* **Note 3**)
17. Phospho-antibodies tested for specificity and titrated, such as the following phospho-antibodies from BD: pSTAT1-Alexa647, pSTAT6-Alexa488, pSTAT3-Alexa488, pSTAT5-Alexa647, CD3-PE, CD20-PerCP-Cy5.5 (*see* **Note 4**)
18. If applicable uncoated hardware such as 6-well plates for in vitro cell activation (*see* **Note 3**)
19. FACS tubes
20. Calibration and compensation beads. For Calibur cytometer use BD Calibrite beads for all four colors (*see* **Note 5**)
21. Flow cytometer equipped with Argon laser (448 nm) and Red Diode laser (635 nm); four fluorescent detectors: FL1: 530±15 nm, FL2: 585±21 nm, FL3: 670 nm, FL4: 661±8 nm. We use a bench top FACSCalibur dual-laser cytometer (BD).
22. Analyzing software for flow cytometry data, such as FlowJo (Treestar)
23. Rotator for incubator, optional

3. Method

Collection and handling of PBMC is described in **Section 3.1**, in vitro activation and phosphostaining protocol in **Section 3.2**, and data analysis in **Section 3.3**.

3.1. Handling of PBMC

1. Prepare: Serum-free media (4°C). ProFreeze media (4°C). Ice bucket. Sterile PBS. Mr. Frosty filled with isopropanol
2. Collect whole blood in BD Vacutainer CPT tubes (*see* **Note 1**)
3. Process within 2 h according to manufacturer instruction
4. Resuspend PBMC in 4°C X-Vivo 15 serum-free media at 15.0×10^6 cells/ml.
5. Work on ice
6. Carefully add chilled ProFreeze-CDM media supplemented with 15% DMSO to equal volume of cell suspension. The final cell concentration will be 7.50×10^6 cells/ml, with 7.5% DMSO.
7. Transfer suspension to 4°C cryovials, place them in Mr. Frosty and leave container at –80°C overnight before transferring to liquid nitrogen for long term storage.

3.2. In Vitro *Activation Staining*

1. Prepare:
   - –20°C : 95% MeOH
   - 37°C water bath:

     16% paraformaldehyde ampoule
     Serum-free support media X-Vivo
     Appropriate number of 9 ml X-Vivo media in 10 ml Falcon tubes to transfer PBMC from cryotubes
2. Optimize stimulation time and concentration of stimulus for your experimental conditions (*see* **Note 3**)
3. Titrate the antibodies if necessary (*see* **Note 4**)
4. Compensation for fluorescent spillover (*see* **Note 5**)
5. Thaw cells quickly in 37°C water bath for 1–2 min, dry cryovials and wipe with alcohol. Open tube in cell culture hood and transfer to 9 ml warm X-Vivo when the cell suspension is still partly frozen. You need $0.5–1.0 \times 10^6$ cells for each stimulation/staining combination
6. Spin the cells gently at 250–300 *g* for about 7–10 min
7. Resuspend cells and add enough serum-free support media for in vitro activation at a cell concentration of $0.5–1.0 \times 10^6$

cells/ml. Add more media to account for pipetting errors. Let the cells equilibrate for 1–2 h in incubator. (*see* **Note 3**)

8. Split cells into Falcon tubes or FACS tubes, one for each stimulus plus 1 for unstimulated cells

   Make sure you have enough cells to split them for staining at 0.5–1.0 × 10^6 cells/ml for every antibody cocktail. Let cells rest for an additional 30 min before activation (*see* **Note 3**)
9. Add cytokines at a concentration of 20 ng/ml and stimulate cells in incubator for 15 min or as optimized in your preliminary experimental setup (*see* **Note 3**)
10. Stop reaction with warm PFA at a final concentration of 1.6%. Mix gently by pipetting. PFA is toxic and volatile. Take appropriate precautions
11. Let cells fix at room temperature for 10 min
12. Spin cells at 500–800 *g* for 7–10 min. Check your steps. Discard supernatant
13. Resuspend cells in small volume of PBS (e.g., 50 μl). This will prevent clumping of cells when MeOH is added. Check cells under the microscope
14. Add ice-cold 95% MeOH. Generally you need about 1 ml MeOH to permeabilize 1 × 10^6 cells. Mix well by gently vortexing. Important: Check for clumping under the microscope. Use a 10 ml pipette if clumping persists. Be patient: cells will un-clump if you resuspended them in PBS before you add MeOH
15. Incubate at room temperature for 10 min
16. To help pelleting of cells in MeOH, add about 1–2 ml of PBS
17. Spin cells at 500–800 *g* for 7–10 min
18. Wash cells with 1–2 ml PBS
19. Resuspend cells in appropriate volume of PBS. You need 50 μl for each staining plus 5–10% to account for pipetting errors
20. Prepare antibody cocktails A, B (*see* **Note 4**)

A:		**B:**	
STAT1-Alexa647	10 μl	STAT5-Alexa647	10 μl
STAT6-Alexa488	10 μl	STAT3-Alexa488	10 μl
CD3-PE	10 μl	CD3-PE	10 μl
CD20-PerCP-Cy5.5	10 μl	CD20-PerCP-Cy5.5	10 μl
Staining Buffer	10 μl	Staining Buffer	10 μl

21. Add appropriate antibody master mix to corresponding FACS tube. Work on ice and protect antibodies from light sources
22. Add 50 μl of cell suspension to appropriate tube, vortex briefly
23. Incubate reaction at room temperature in the dark for 30 min
24. Add 2 ml staining buffer for first wash. Spin 300–500 *g* for 5–10 min
25. Resuspend and repeat step 23
26. Resuspend cells in 200–250 μl of Staining Buffer. If cells are too diluted you will spend a lot of time on one sample to collect enough cells on rare cell subtypes
27. Keep cells cool and dark until flow analysis, for example, Styrofoam box with ice and lid.

3.3. Data Acquisition

1. Calibrate cytometer following instruction for your particular machine, for example, Calibrite beads
2. Optimize instrument settings with your own cell samples (*see* **Note 5**)
3. If possible, set up instrument to collect enough events for the least abundant cell subtype, the more the better!

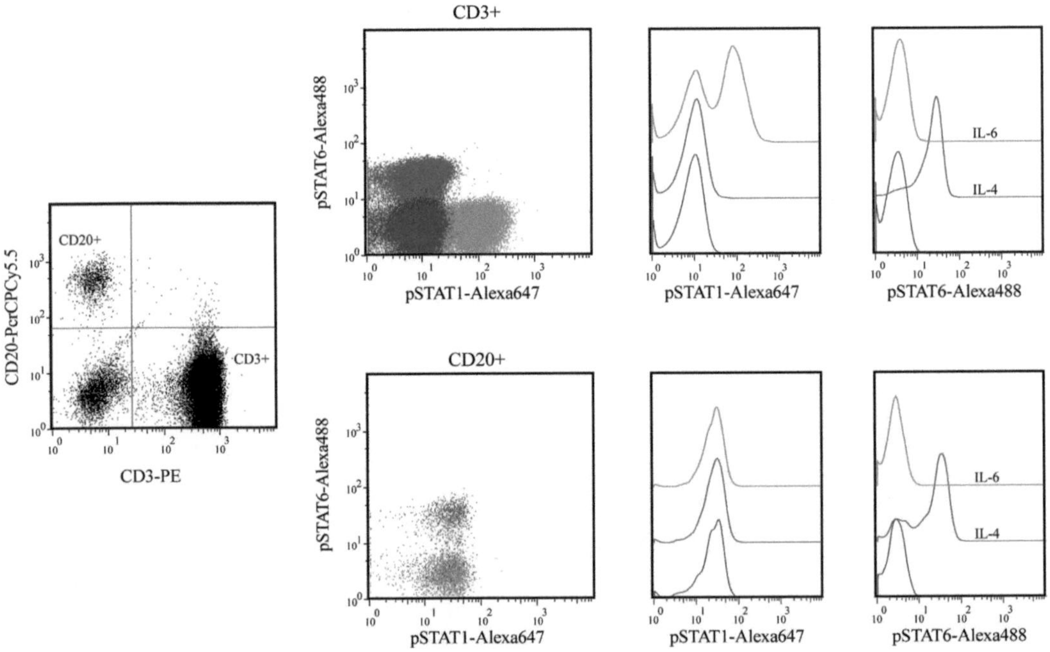

Fig. 4.1. Multidimensional analysis of human PBMC stimulated with either IL-6 (*green*), IL-4 (*red*) or *left* untreated (*blue*). Cells were fixed and permeabilized following protocol 3.2 and stained simultaneously with antibody cocktail A. *Top panels* show superimposed *dot plots* and *histograms* for T-cells ($CD3^+$), the *bottom panels* show B-cells ($CD20^+$). In overlays the induction of specific phosphorylation events are clearly identifiable. (*See* Color Plate 1)

4. Carefully resuspend cells by vortexing. To obtain data on a per-cell basis it is essential to avoid clumps.

3.4. Data Analysis

1. Export data to FlowJo or similar program
2. Gate cells of interest according to cell markers, **Fig.** 4.1 and Color Plate 1
3. Make overlay histograms of unstimulated versus stimulated samples for every cytokine used, **Fig.** 4.2 and Color Plate 2

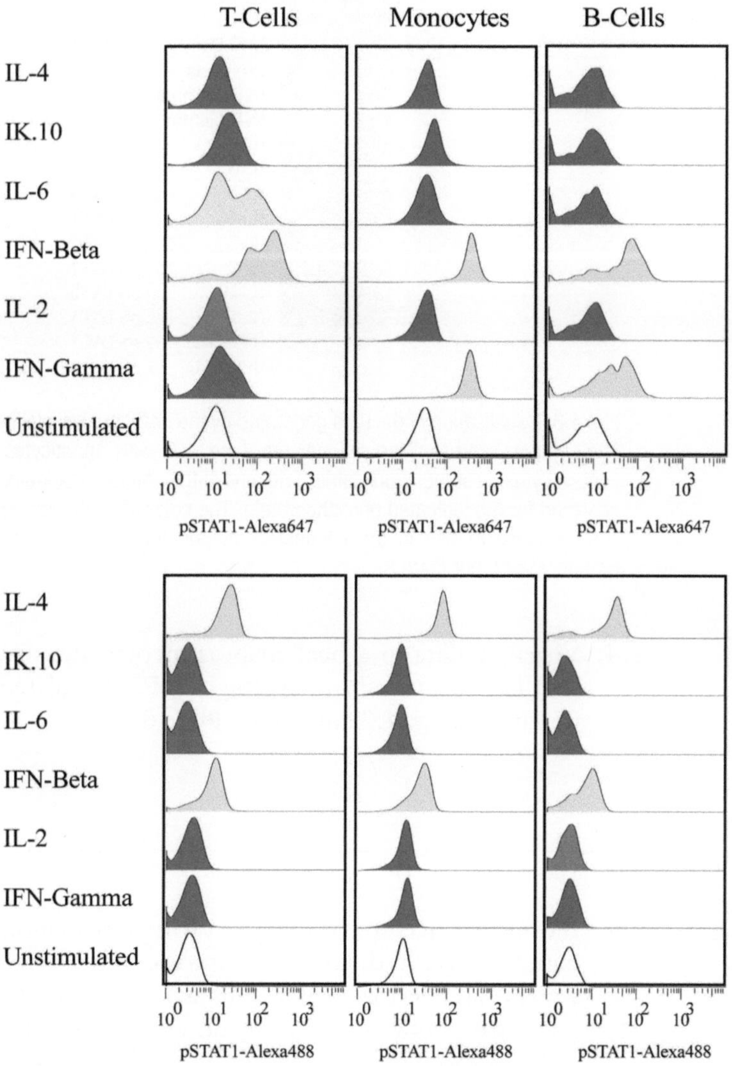

Fig. 4.2. PBMC were stimulated with indicated cytokines, fixed and permeabilized according to protocol 3.2. T-cells ($CD3^+$) and B-cells ($CD20^+$) were gated according to markers while monocytes were gated in *scatter plot. Open histograms* represent untreated cells, *filled histograms* stimulated cells. Induction of phosphorylation is clearly identifiable (*filled yellow histograms*). (*See* Color Plate 2)

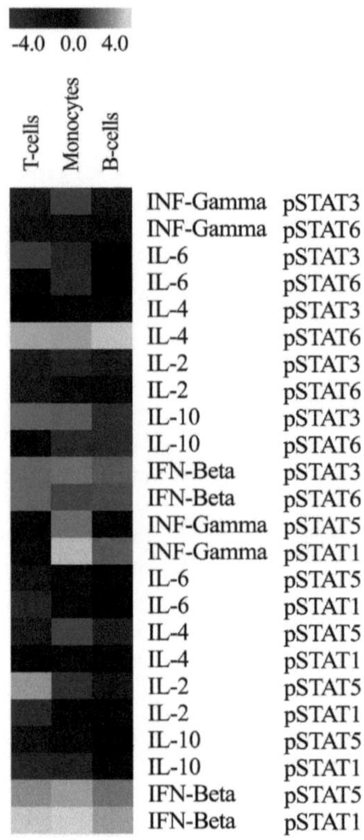

Fig. 4.3. Visualization of the data generated by the FACS analysis following protocol 3.2. The columns represent the cell subsets, T-cells, B-cells, monocytes. Each row represents a cytokine stimulation stained with one of the antibody cocktails and subsequently analyzed for the indicated phosphoprotein. The color of each block represents the fold change ($\log_2$) in MFI in the channel corresponding to the analyzed phophorylated protein. (*See* Color Plate 3)

4. Visualize data in a heat map as median fluorescent intensity (MFI), $\log_2$ transformed: $\log_2$ (MFI stimulated/MFI unstimulated), **Fig.** 4.3 and Color Plate 3

4. Notes

1. Blood cells need to be collected with a meticulous attention to handling. Method, time, temperature, and storage can affect cell viability and performance of this assay *(12–15)*. BD CPT –Vacutainer tubes offer sterile collection of peripheral blood which is stable at room temperature for up to 24 h. This offers an off-site collection possibility. The tubes are available in different sizes. Make sure you have fitting adaptors for swinging bucket rotors. Cells can also be collected with the traditionally used Ficoll Hypaque method. Serum-free support media has been used

for both cryopreservation and in vitro activation. We have been testing signaling in both serum free media and autologous sera. The basic protocol works for sera as well, but signaling needs to be evaluated for every new assay.

Cryopreseravtion of cells in general is a very delicate step, especially for rare cells. Cell density in cryovials can vary but in our experience it is of interest to work consistently. If cell viability is lower then expected check this step carefully since not all cells react the same to cryopreservation. Depending on populations under investigation it may be beneficial to include sera for support.

2. Staining Buffer: Non-specific staining can be higher in fixed and permeabilized cells. It is recommended to add extra protein like BSA to the staining buffer. Sodium azide can be added to inhibit microbial growth. By making fresh staining buffer for each run one can avoid the use of toxic sodium azide.

   Paraformaldehyde solution, methanol free, is an efficient and rapid penetrating fixative that works particularly well in this assay to denature and thereby preserve the phosphorylation state of proteins within cells for later staining.

   Methanol is an efficient agent for the permeabilization of PBMC. It has been used at different concentration of 70–100% *(6)* and is particularly suited for nuclear antigens. Saponin is a further reagent used for permeabilization, typically at concentrations of 0.1–0.5%. Some antigens are preserved better with saponin *(9)*.

3. For in vitro stimulation it is possible to use compounds such as PMA or recombinant peptides to activate the signaling pathway of interest. Careful preparation of the intended analysis will help in the decision process of which and how many stimuli to use in a particular assay and which and how many subpopulations to study at first. Generally, if ones intention is to study a particular pathway it is important to choose a compound that strongly activates the signaling cascade studied. If one is interested in evaluating the effect of a drug it may be of interest to study more than one pathway simultaneously.

   In this protocol the phosphorylation of various STAT molecules in response to a panel of cytokines is being assessed. Cytokines are usually shipped lyophilized. Reconstitute them according to manufacturer instruction and make sure they have been cell cultured tested. To avoid repetitive freeze/thawing aliquot the solutions. You may want to test the reconstitution media for signaling inhibition by stimulating one sample with just reconstitution media.

   Check the literature for a starting point on concentration of stimulus used in cell culture. To find an appropriate stimulation time for your experimental conditions titer the stimulating agent for your cell concentration and conditions. If you use multiple stimuli in one run you may want to compromise and find a stimulation time that activates all the molecules of interest sufficiently. In this particular protocol 15 min are appropriate for all the cytokines. Perform a serial dilution to find optimal stimulus concentration for your assay. Take special care to work consistently. Cell density and stimulation time affect the outcome of this assay and need to be monitored accordingly, especially if the analysis of the data is of a quantitative nature. Working with antibody master mixes when applicable will increase the accuracy.

Hardware must be tested since some cell types tend to stick to plastic, especially if they are activated. Use uncoated hardware and rotator if necessary. Check that the cells are in suspension before any transfer. Cell loss can be quite substantial. We have used Falcon tubes on a rotator or uncoated plates for stimulation Depending on the pathway and cell type studied you may want to let the cells rest for up to 2 h. Cells need to equilibrate and longer resting periods can help cells get back to a basal phosphorylation state after cryopreservation. This will facilitate better activation.

It is advisable to practice the procedure on cell lines, but keep in mind that cell lines are not primary cells and that their signaling can differ significantly

4. The antibody–fluorophore conjugates used in this assay are tested for their phospho-specificity and their suitability for flow cytometry. Importantly, all antibody–fluorophore conjugates withstand paraformaldehyde fixation and methanol permeabilization. We have titrated the antibodies for this assay and found that 10 μl cell surface makers were enough to be able to easily determine the CD3 and CD20 positive populations. Titration of the intracellular phospho-antibodies in unstimulated and stimulated samples showed that 10 μl gave a good signal to noise ratio. It is recommended to titrate the antibodies for every new assay to insure optimal staining for phenotypic markers and phospho-antigens under specific experimental conditions. The links provided below are good starting points to plan multicolor experiments. Pay attention to which kind of permeabilization reagent has been tested for a particular antibody. Commonly it will be MeOH or saponin.

   If the antibody–fluorophore conjugates are not available commercially, you will need to run appropriate tests for your antibody and fluorophore. Antibody specificity must be checked by western blot. Test different clones to find those that recognize single bands. Keep in mind that antibodies to phosphoproteins are mostly raised against short peptides in denatured form. Use inhibitors or peptide competitors of the signaling cascade to

demonstrate activation through phosphorylation by the signaling molecule tested. For the choice of fluorophore you need to consider the wavelengths of the lasers and the filters. Use small and photostable fluorophores especially for intracellular staining, such as the alexa dyes. Choose bright fluorophores for rare antigens. Dyes and antigens are susceptible to fixation and permeabilization and need to be evaluated for your particular assay. Fortunately, companies are starting to provide information on suitability of a particular phospho-antibody conjugate with certain fixation and permeabilization protocols, see links. Fluorophore conjugation protocols are available from the manufacturers themselves. For example, see Invitrogen for the highly stable alexa dyes.
http://www.bdbiosciences.com/docs/Validated_Cell_ Surface_Markers.xls
http://beckmancoulter.com/literature/Bioresearch/ISAC2004_FMalergue.pdf

5. For multicolor compensation you need single stained cells or compensation beads for each antibody–fluorophore conjugate. Keep in mind that monocytes and lymphocytes do not have the same auto-fluorescence. Compensate lymphocytes within the lymphocyte gate and monocytes within the monocyte gate. Modern cytometer software can do the compensation for you. To perform a proper compensation you need cells that are brightly stained and cells that are unstained or slightly stained within the same tube. If, for example, all PBMC are activated by a certain compound you can add unstimulated cells to the same tube. Get help from a knowledgeable person if this is your first attempt. You can find a comprehensive discussion on the following webpage: http://www. drmr.com/compensation/indexDetail.html

   For relative fluorescence intensity measurements your cytometer needs to be calibrated every time to account for instrument variability. Follow the manufacturer's instructions for calibration. For this four-color flow cytometer protocol you need Calibrite 3 Beads Kit with PerCp-Cy5.5 and Calibrite APC and software FACSComp. (If you use a different cytometer use the manufacturer's instruction for calibration and compensation.). Optimize the instrument settings with your own cell samples. Adjust forward and side scatter in order to easily distinguish lymphocytes from monocytes. Adjust compensation for fluorescent spill-over using the single stained tubes. The compensation settings can be stored and used in subsequent assays. We compensated the two alexa dyes in the lymphocyte gate and checked if the compensation worked for the monocytes, which it did. Set up the cytometer so that sufficient events will be collected for the least abundant population. To get distinct peaks you need to collect a sufficient number of events. The larger the events the more distinct the peaks will be. In this case, we determined that a minimum of 1,000 events for CD20 B-cells was sufficient to provide analyzable data.

References

1. Perez OD, Nolan GP. Phospho-proteomic immune analysis by flow cytometry: from mechanism to translational medicine at the single-cell level. Immunol Rev 2006; 210:208–28.
2. Hunter T. Signaling–2000 and beyond. Cell 2000;100:113–27.
3. Mustelin T, Vang T, Bottini N. Protein tyrosine phosphatases and the immune response. Nat Rev Immunol 2005;5:43–57.
4. Bayer J, Grunwald D, Lambert C, Mayol JF, Maynadie M. Thematic workshop on fluorescence compensation settings in multicolor flow cytometry. Cytometry B Clin Cytom 2007;72:8–13.
5. Perez OD, Mitchell D, Nolan GP. Differential role of ICAM ligands in determination of human memory T cell differentiation. BMC Immunol 2007;8:2.
6. Montag DT, Lotze MT. Successful simultaneous measurement of cell membrane and cytokine induced phosphorylation pathways [CIPP] in human peripheral blood mononuclear cells. J Immunol Methods 2006; 313:48–60.
7. Krutzik PO, Clutter MR, Nolan GP. Coordinate analysis of murine immune cell surface markers and intracellular phosphoproteins by flow cytometry. J Immunol 2005;175: 2357–65.
8. Krutzik PO, Hale MB, Nolan GP. Characterization of the murine immunological signaling network with phosphospecific flow cytometry. J Immunol 2005;175: 2366–73.
9. Perez ODOD, Krutzik POPO, Nolan GPGP. Flow cytometric analysis of kinase signaling cascades. Methods Mol Biol 2004;263:67–94.
10. Krutzik PO, Irish JM, Nolan GP, Perez OD. Analysis of protein phosphorylation and cellular signaling events by flow cytometry: techniques and clinical applications. Clin Immunol 2004;110:206–21.
11. Krutzik PO, Nolan GP. Intracellular phospho-protein staining techniques for flow cytometry: monitoring single cell signaling events. Cytometry A 2003;55:61–70.
12. Weinberg A, Betensky RA, Zhang L, Ray G. Effect of shipment, storage, anticoagulant, and cell separation on lymphocyte proliferation assays for human immunodeficiency virus-infected patients. Clin Diagn Lab Immunol 1998;5:804–7.
13. Debey S, Schoenbeck U, Hellmich M, et al. Comparison of different isolation techniques prior gene expression profiling of blood derived cells: impact on physiological responses, on overall expression and the role of different cell types. Pharmacogenomics J 2004;4:193–207.
14. Maecker HT, Rinfret A, D'Souza P, et al. Standardization of cytokine flow cytometry assays. BMC Immunol 2005;6:13.
15. Ruitenberg JJ, Mulder CB, Maino VC, Landay AL, Ghanekar SA. VACUTAINER CPT and Ficoll density gradient separation perform equivalently in maintaining the quality and function of PBMC from HIV seropositive blood samples. BMC Immunol 2006;7:11.

Chapter 5

Investigating T Cells by Polychromatic Flow Cytometry

Enrico Lugli, Leonarda Troiano, and Andrea Cossarizza

Abstract

Since its development, flow cytometry gave a relevant contribution to the field of Immunology. Its unique potential to analyse multiple parameters at the single cell level allowed the identification of unknown cell subsets with specific roles in immunoregulation as well as in the pathogenesis of several diseases. More recently, with the advent of new equipments and fluorochromes, the possibility exists to analyse simultaneously a large number (up to 19) of parameters in a single cell. This strategy, defined polychromatic flow cytometry (PFC), has been widely utilised in the last years for the fine analysis of immune cell phenotypes, including antigen-specific T lymphocytes, B cell subsets, and the intracellular phosphoproteome, among others. A huge amount of data can be generated by such an approach, and their interpretation could become a very complex and time-consuming task. Protocols for performing PFC will be discussed in this chapter, together with some guidelines for data interpretation and analysis.

Key words: Polychromatic flow cytometry, monoclonal antibodies, fluorescence compensation, single stained controls, fluorescence-minus-one controls, surface antigens, intracellular antigens, data analysis, cluster analysis.

1. Introduction

Fluorescence-activated flow cytometry is the powerful technology that allows the analysis of the fluorescence emitted by cells in suspension that have been previously stained with fluorescent probes or fluorochrome-conjugated monoclonal antibodies (mAbs). Cells are conveyed in the flow chamber through a capillary which ensures the transition on a cell-by-cell basis and, thus, allows single cell analysis of light scatter and fluorescent signals. At the end of 1960s, the first commercial flow cytometer, i.e. the "Impulsecytophotometer ICP-11", developed in 1968 by Dr. Wolfgang Göhde/Partec and distributed by Phywe AG,

Gennaro De Libero (ed.), *T Cell Protocols: Second Edition, vol. 514*

DOI 10.1007/978-1-60327-527-9_5 Springerprotocols.com

Göttingen, had optionally one or two fluorescence parameters for the analysis of DNA content and ploidy. Scatter parameters were not featured by this instrument.

The technology knew a rapid growth during the next years, that became exponential in the era of HIV infection. Indeed, at the end of 1980s tools capable of measuring up to four colours were developed. In the last years, flow cytometry has been characterised by further improvements, that are giving the possibility to measure simultaneously up to 19 parameters in the same cell *(1, 2)*. However, along with the increase in the number of parameters that can be measured, experiments and data analysis are becoming more complex. Post-acquisition computer assistance is required to minimise possible errors thanks to the generation of a compensation matrix, and to reduce the complexity of huge datasets generated by such a technology. We will briefly illustrate the steps that we follow for an adequate use of polychromatic flow cytometry (PFC), and indicate how to perform analysis of these data either by conventional or new, original approaches.

2. Materials

1. Fresh (better!) or cryopreserved peripheral blood mononuclear cells (PBMCs);
2. Phosphate buffer saline (PBS), store at 4°C;
3. Foetal Bovine Serum (FBS), store at –20°C;
4. Staining buffer: PBS + 0.5% Bovine Serum Albumin (BSA), store at 4°C; an alternative staining buffer can be used in order to avoid cell loss during centrifugation: PBS + 5% FBS, which can be stored at 4°C for up to 1 month;
5. Fluorochrome-conjugated monoclonal antibodies;
6. BD™ CompBeads (BD Biosciences, Franklin Lakes, NJ, USA);
7. Viability dye, such as propidium iodide (PI), stock solution at the concentration of 50 μg/mL in PBS, store at 4°C, stable for years;
8. "Monomeric cyanine nucleic acid stains" dyes (Invitrogen Corp. Carlsbad, CA, USA), such as TO-PRO3, stock solution at 1 mM in DMSO, store at –20°C, stable for up to 1 year.
9. Fixable viability dyes, such as "LIVE/DEAD fixable dead cell stain kits" (Invitrogen); once dissolved in DMSO, they are stable for 15 days at –20°C. Caution: DMSO is toxic; avoid ingestion and inhalation or contact with eyes and skin;

10. Paraformaldehyde (PFA), 1% in PBS, pH 7.2, or an equivalent fixation buffer. PFA (1%) in PBS can be stored at 4°C for up to 1 month. Caution: PFA is irritant; avoid ingestion and inhalation or contact with eyes and skin;
11. Tween 20 (0.2%) in PBS, pH 7.2, or an equivalent permeabilisation buffer. Tween 20 (0.2%) in PBS can be stored at 4°C for up to 1 month. Caution: Tween 20 is toxic; avoid ingestion and inhalation or contact with eyes and skin;
12. Three laser (405, 488 and 635 nm) flow cytometer;
13. Software for compensation and data analysis;
14. Softwares for cluster analysis, such as Cluster and Treeview softwares (downloadable at rana.lbl.gov) or TM4 MeV (downloadable at www.tm4.org/mev.html) or similar

3. Methods

The methods described here below refer to the possibility of performing a PFC analysis on peripheral blood lymphocytes. We will briefly describe: (1) the set up of the flow cytometer; (2) how to choose the right fluorochromes; (3) the procedure to correctly compensate fluorescence spillover among different channels; (4) some tricks to avoid the detection of unspecific, false positive events; (5) the protocol for a simultaneous detection of T cell surface markers along with intracellular Ki-67 expression by fluorochrome-labelled mAbs; and (6) how to perform data analysis by classical and recently developed methodologies.

3.1. Set Up of the Flow Cytometer

Currently, the most sophisticated flow cytometers available on the market are already equipped with filters and dichroic mirrors for the detection of several fluorochromes. **Table 5.1** shows the most common fluorochromes employed in the polychromatic analysis of lymphocyte phenotype, the laser line used for their excitation and the suggested wavelength of filters for their detection. Multiparameter flow cytometers that can detect eight colours or more are generally equipped with a blue laser that has a wavelength of 488 nm, a 635 nm red laser and a 405 nm violet laser; other lasers could also be used, such as a 350 nm UV laser for better excitation of fluorochromes such as Hoechst 33342 (for DNA content), monobromobimane (for intracellular reduced glutathione) and Indo-1 (for quantification of intracellular free or protein-bound calcium), which are often used for functional studies of T cells, or a 532 nm green laser for better excitation of propidium iodide, PE-tandems or merocyanine 540.

Table 5.1
Fluorochromes that are commonly used in polychromatic flow cytometry, with their excitation wavelength (λ) and suggested filter used for their detection

Laser excitation	Fluorochrome (Abbreviation)	λ of the filter used for detection (expressed in nm)
488 nm (blue laser)	Fluorescein (FITC)	510–530
	Alexa 488	510–530
	Phycoerythrin (PE)	565–590
	Phycoerythrin-Texas Red (PE-TxR, ECD)	600–620
	Phycoerythrin-cyanine 5 (PE-Cy5, or Tricolour, TC)	650–680
	Perchlorophilin (PerCP)	650–680
	Phycoerythrin-cyanine 5.5 (PE-Cy5.5)	690–730
	Perchlorophilin-cyanine 5.5 (PerCP-Cy5.5)	690–730
	Phycoerythrin-cyanine 7 (PE-Cy7, PC7)	750 long pass
633 nm (red laser)	Allophycocyanine (APC)	650–670
	Alexa 647	650–670
	Allophycocyanine-cyanin 5.5 (APC-Cy5.5)	690–730
	Alexa 700	690–730
	Allophycocyanine–cyanin 7 (APC-Cy7)	750 long pass
	Allophycocyanine-Alexa750 (APC-Alexa750)	730 long pass
404 nm (violet laser)	Cascade Blue (CB)	435–475
	Pacific Blue (PB)	435–475
	Alexa 405	435–475
	Cascade Yellow (CY)	530–570
	Pacific Orange (PO)	530–570

The choice of filters and mirrors is aimed both to detect the brightest signal of the fluorochrome and to minimise the fluorescence compensation among different channels. Different types of filters and mirrors are typically present in a flow cytometer. Dichroic mirrors are able to reflect the fluorescence below a given wavelength and to transmit the fluorescence with a higher wavelength, or vice versa. In general, dichroic mirrors have a wavelength that is in the middle of two strategical fluorescence channels. For example, for the detection of FITC (maximum emission at 520 nm) an PE (maximum emission at 590 nm), a 560 nm dichroic mirror should be adopted. The sequential introduction of dichroic mirrors in the optical setting allows an optimal separation and detection of multiple fluorochromes at the same time. Since in a polychromatic configuration the light arriving from fluorochromes to photomultipliers (PMT) is selected through several mirrors, it is extremely important to check carefully the transmission power of mirrors in order to avoid loss of fluorescence *(3)*.

Filters in front of the PMT are equally important and further collect the light transmitted or reflected by dichroics. These filters can be either band-pass (collecting the light between two specific wavelengths) or long-pass (collecting the light higher than a specific wavelength). The choice of filters depends upon the emission wavelength of the fluorochrome that has to be measured. These filters should be wide enough to collect the maximum light arriving from the excited fluorochrome (**Table 5.1**).

Note: Most flow cytometers now have a fixed configuration and there is no need to work on the hardware. However, in several instruments the possibility exists to change the configuration according to the requirement of the user.

3.2. Choose the Right Fluorochrome

A variety of fluorochromes can be used to detect antigens by PFC, and thus thousands of combinations are literally possible. However, not all fluorochromes are suitable for the detection of certain antigens. Indeed, there are two main aspects that should be considered when choosing a particular antibody-conjugate: (i) the brightness of the fluorochrome and the relative abundance of the antigen of interest; (ii) the requirements of the fluorescence compensation among different channels.

3.2.1. Brightness of the Fluorochrome and Amount of Antigen

Not all the fluorochromes have the same brightness; as a consequence, those brighter than others, such as PE, PE-Cy5, PE-Cy7 and APC, should be adopted for the detection of antigens which have low expression (such as cytokines, chemokines and their receptors), or for rare events (intracellular cytokines in antigen-specific assays, or intracellular Ki-67 in unstimulated lymphocytes). For example, only PE-conjugated mAbs are suitable for the detection of the expression of CD127 (the IL-7 receptor α-chain) on CD3+ T cells, while APC-conjugated mAbs are not.

It is suggested to adopt APC-tandem dyes for the detection of well-expressed lineage antigens, i.e. CD3, CD4, CD8, where a good and clear separation exists between negative and positive populations. Be careful with certain violet dyes, mainly with Cascade Blue or similar (Pacific Blue and Alexa 405) which can bind in a non-specific manner to dead cells *(4)* and Lugli et al., unpublished observation). This implies that these dyes can be used not only for the detection of certain markers but also provide a good tool for excluding dead cells from the analysis *(4)*. Finally, take into account that tandem fluorochromes tend to degrade faster (the acceptor fluorochrome can detach from the donor and thus the signal is lost), sometimes before the expiration date indicated in the label.

3.2.2. Compensation Requirements Among Different Channels

It is important to consider the interference of other fluorochromes into the channel of interest (*see* **Section 3.3**). In fact, the dyes or probes used in PFC can have wide emission spectra, and their fluorescence can be detected in different channels *(5)*. In some cases, this interference is relevant, and thus the detection of certain antigens is almost impossible. This is the case for the probe Carboxyfluorescein Diacetate-Succinimidyl Ester (CFSE), commonly used to detect clonal proliferation of T cells. CFSE, which is typically detected in the FITC channel, has a very broad emission spectrum and displays a considerable spillover into PE and PE-Cy5 channels. Hence, it is very difficult to detect antigens with a low expression in these channels, even if antibodies conjugated to bright fluorochromes are used. **Figure 5.1** shows that PE anti-CD127 mAb is not suitable for the detection of CD127 expression when used with CFSE. In fact, in the case shown here, 57% of peripheral blood lymphocytes express CD127; however, in cells co-stained with CFSE (and after proper compensation), CD127+ cells are mostly undetectable. Thus, in this case, it is recommended to analyse CD127 (or antigens with low expression) in other channels, where the spillover of CFSE is minimal, such as in channels collecting light from violet fluorochromes. PE and PE-tandems can be utilised, but only for recognising brighter antigens, such as lineage molecules such as CD3, without a relevant loss of sensitivity.

3.3. Fluorescence Compensation in PFC

Fluorochromes commonly used in polychromatic experiments display spectral overlap *(5)*. As a consequence, a particular fluorochrome can be measured in multiple PMTs. These signals are to be eliminated by a process called compensation, which is able to mathematically subtract the unwanted fluorescence in a specific detector *(6)*. In PFC, compensation requirements are very complex because the emission of each fluorochrome is to be subtracted from all the channels and the number of pair-wise

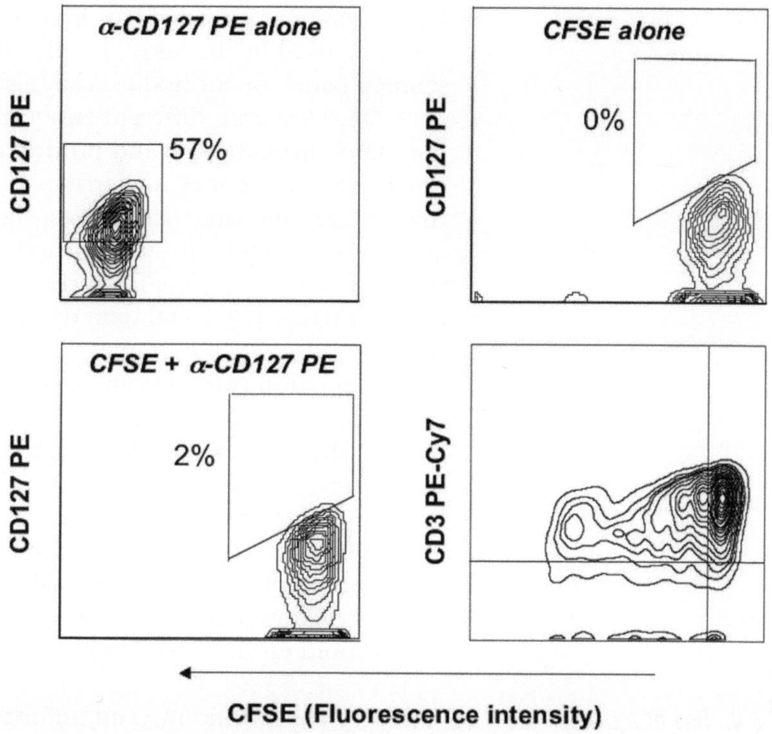

Fig. 5.1. Effect of the fluorescence spillover on the detection of antigens with low expression.
PBMCs were stained either with anti-CD127 PE, CFSE or both, then samples were compensated with FlowJo 6.3 for MacOSX. Percentages are given for CD127+ lymphocytes within the gate. In *lower right panel*, the staining with CFSE and PE-Cy7 CD3 is shown for a comparison. Note that CD3+ cells are easily recognisable.

combinations grows exponentially with the number of parameters measured. Nowadays, compensation can be automatically performed by dedicated softwares which are able to simplify the procedure and minimise errors. In order to compensate correctly, samples stained with single fluorochromes are prepared. When compensating, consider the following aspects:

- For single stained samples, employ either the cells that will be used in the experiment or CompBeads™. CompBeads™ are beads which are able to bind light chain-bearing immunoglobulins, display a very high fluorescence in the channel of interest and provide both positive and negative signals which can be used to perform multicolour compensation. Since the antibodies used for compensation are the same as the ones used for the assays, artefacts due to reagents variability are avoided.
- When performing compensation, always gate in the population of interest. In fact, different autofluorescence values, which generally characterise different cell populations (such as lymphocytes and monocytes), can influence compensation.

- Use a compensation control that is at least as bright as the antibody used in the assay, i.e. the antibodies chosen for the staining panel, or antibodies recognising a very bright antigen, i.e. CD8. However, different antibodies, even if conjugated to the same fluorochrome and produced by the same company, can display different characteristics (see below).
- Always use the same antibodies chosen for the staining of interest. Be careful especially to tandem dyes which can behave very differently not only from company to company, but also from lot to lot *(7)*. In tandem dyes, two molecules are chemically conjugated (for example, PE to Cy5), but the efficiency of the conjugation reaction can be very variable, and, thus, affect compensation in a different way. For this reason, compensation matrix should always be checked when using different conjugates.
- Be always sure that your instrument has a correct laser alignment. Check this by running rainbow beads and ensure that the performance of the cytometer is the same day by day *(3)*. An improper alignment of the lasers and alterations in their time delays could influence compensation.

3.4. Use of a Dump Channel to Exclude Unspecific Events

3.4.1. Exclusion of Dead Cells

Frozen cells are commonly (and unfortunately) utilised in a large amount of immunological researches. Freezing and thawing procedures cause cell rupture and death, often relevant. When performing multicolour experiments starting from frozen cells, the following aspects should be considered: (i) antibodies like dead cells and debris, leading to unspecific positive events; (ii) certain antigens can be altered on dead cells, and false negative/dim cell population can appear in the population of interest; (iii) dead cells are characterised by a higher autofluorescence and can generate false positive signals, or can affect compensation. The inclusion of dead cells in the analysis is particularly harmful when analysing rare subsets. For this reason, a dye able to track dead cells is to be included (**Fig. 5.2**). When performing surface immunophenotyping, dyes such as PI (excited by a 488 nm laser and emitting in the PE-TxR channel) or similar (i.e. monomeric cyanine nucleic acid stains) can be used.

1. Resuspend cells in staining buffer;
2. Stain cell surface antigens with fluorescent-labelled mAbs (see below);
3. Wash with 2 mL of staining buffer at 300 *g* for 5 min;
4. Resuspend cells in staining buffer;
5. Add 1 μg/mL PI or 2 nM TO-PRO3 or equivalent dye to the cell suspension, mix thoroughly and analyse on cytometer;
6. Exclude dead cells by gating PI or TO-PRO3 negative cells on SSC vs PI/TO-PRO3 dot plot;

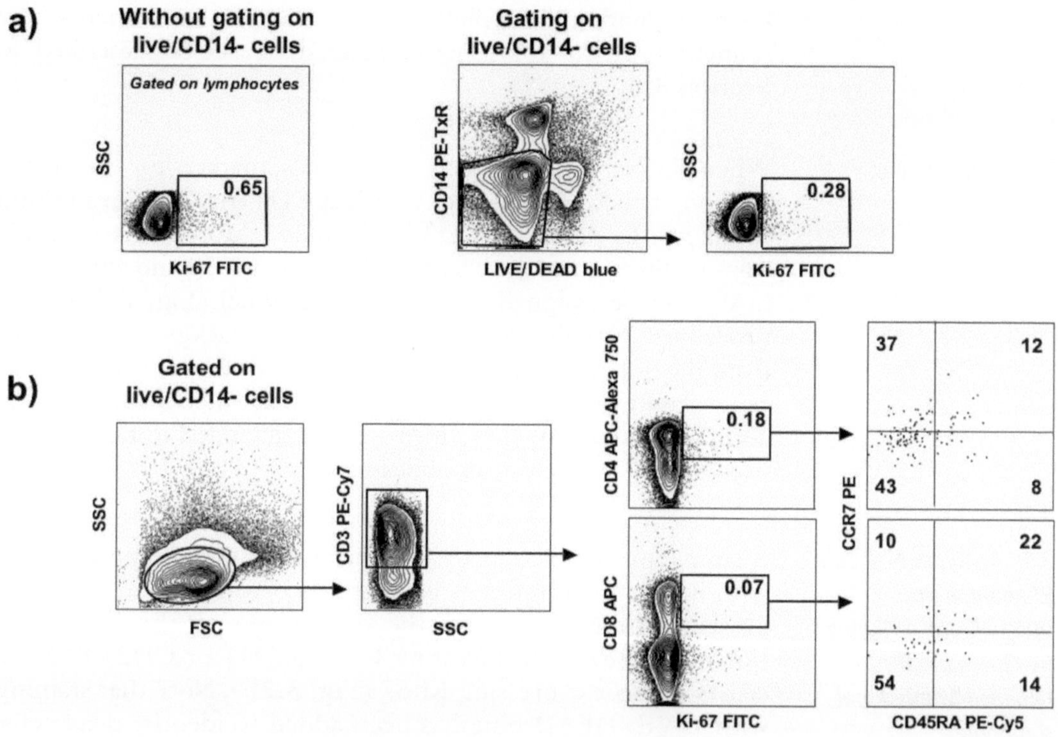

Fig. 5.2. Analysis of Ki-67+ T cells by polychromatic flow cytometry.
PBMCs were simultaneously stained with LIVE/DEAD blue to identify dead cells and several mAbs to quantify the expression of CD3, CD4, CD8, CD14, Ki-67, CD45RA and CCR7. (**a**) The presence of dead cells can influence the analysis of rare subsets. The percentage of Ki-67+ lymphocytes before (*left*) and after (*right*) gating on live/CD14- cells is indicated. Note that without the gate on CD14- living cells the percentage of false positive Ki-67+ cells is more than double. (**b**) Further analysis of the characteristics of Ki-67+ T cells. Dead cells and monocytes were excluded as in (a), then lymphocytes were identified by gating on forward (FSC) and side scatter (SSC), and T cells on the basis of CD3 expression; helper and cytotoxic subsets of T cells were then selected by CD4 and CD8 positivity, respectively. The percentages of proliferating Ki-67+ CD4+ and CD8+ T cells are indicated. Inside Ki-67+ cells, further analysis of CD45RA and CCR7, which define T cell differentiation status, are shown. Numbers indicate the percentage of cells inside the gate.

Generally, propidium iodide is chosen. Also, TO-PRO 3 works very well in our hands; however, since this dye is excited by the 635 nm red laser and emits around 660 nm, it is not possible to include APC-conjugated antibodies in the staining.

Differently, when performing analysis on permeabilised samples, it is necessary to stain dead cells before the permeabilisation step, in order to avoid the staining of the whole cell population. Traditionally, ethidium monoazide (EMA), a fixable dye that enters dead cells and covalently links to DNA after light exposure, has been used for this purpose. However, recently, the LIVE/DEAD family dyes have become available: they are ready to use, display very high fluorescence and can be excited with different lasers, facilitating the application in multicolour experiments. The staining protocol for LIVE/DEAD dyes will be described in **Section 3.5**.

3.4.2. Exclusion of Non-T Cells

When performing PFC analysis of T cells, markers for the identification of major T cell subsets, such as CD4 and CD8, in addition to the *pan* T cell marker CD3, are included. Moreover, to exclude non-T cells from analysis, anti-CD19, anti-CD56 and anti-CD14 mAbs can be assigned to a "dump" channel (**Fig. 5.2a**). This procedure partially reduces fluorescence background derived from the non-specific binding of mAbs to cell types other than T cells, mainly monocytes. In order to save another channel for the antigens of interest, these mAbs can be included in the channel used for the identification of dead cells.

3.5. Analysis of T Cell Phenotype and Intracellular Antigens by Fluorescent-Labelled Monoclonal Antibodies

A simple protocol to analyse T cell surface phenotype together with intracellular antigens is described below. In particular, we present a 8-colour assay for the identification of the differentiation status *(8)* of proliferating CD3+,CD4+ or CD3+,CD8+T cells, i.e. those expressing Ki-67 (**Fig. 5.2b**). Note that staining with LIVE/DEAD blue has been added to identify dead cells, and anti-CD14 mAb to exclude monocytes from the gate of analysis.

1. Resuspend 1×10^6 peripheral blood mononuclear cells (PBMCs, previously isolated by Ficoll density gradient centrifugation) in 100 μL of staining buffer;
2. Add mAbs for surface markers at the pre-titrated concentrations. Always titrate antibodies before use. For detailed antibody titration procedures consult the paper by Kantor and Roederer *(9)*;
3. Incubate for 20 min at room temperature (RT). Note that some antigens, such as chemokine receptors, are better recognised by incubation with mAbs at 37°C;
4. Wash at 300 *g* for 5 min at RT with staining buffer. In order to avoid cell loss, washing steps can be performed with PBS + 10% FBS;

Note: It is possible to avoid dead cell staining (steps 5–10) if working with freshly isolated cells

5. Wash with PBS;
6. Resuspend the cells in 1 mL of PBS;
7. Adjust the cell density to 1×10^6 cells/mL;
8. Add 1 μL/10^6 cells of the LIVE/DEAD fixable dead cell stain solution;
9. Incubate for 30 min at room temperature (RT);

10. Wash with PBS;
11. Resuspend cells in 1 mL of PBS + 1% PFA or equivalent fixation buffer;
12. Incubate 15 min at RT;
13. Wash with staining buffer;
14. Wash with 2 mL of PBS + 0.2% tween 20 or equivalent permeabilisation buffer;
15. Resuspend cells in 100 μL of PBS + 0.2% tween 20 or equivalent permeabilisation buffer;
16. Add anti-Ki67 monoclonal antibody;
17. Mix thoroughly and incubate for 30 min at 4°C;
18. Wash with staining buffer;
19. Resuspend cells in 1 mL of PBS + 1% PFA;
20. Prepare compensation controls using single-stained compensation controls or CompBeads™;
21. Acquire compensation controls on flow cytometer;
22. Analyse samples on flow cytometer within 24 h;
23. Acquire at least 300,000 events;
24. Create compensation matrix by software;
25. Compensate experimental samples;
26. Analyse samples as described below.

3.6. Data Analysis

3.6.1. How to Set the Gates: "Fluorescence Minus One" Controls

During the analysis of the data, it is obviously crucial to distinguish the presence or not of the antigen of interest. This is very simple when investigating antigens whose positive expression is clearly distinguishable from the background fluorescence, i.e. CD3, CD4, CD5, CD8, etc. However, in most cases, antigens with unknown patterns of expression are investigated, and it is difficult to set the threshold of positivity. Unstained samples, or those incubated with irrelevant mAbs of the same isotype and colour, are absolutely not reliable for this purpose. "Fluorescence minus one" (FMO) controls must be utilised *(1)*. FMO controls are samples that have been stained with all antibodies except the one of interest. When performing multicolour experiments, in addition to single stained samples, prepare FMO controls for all the antigens under investigation, then set the gates to distinguish positive and negative expression and analyse data as described below.

3.6.2. Conventional Methods for the Analysis of Flow Cytometric Data

The most common method for the analysis of flow cytometric data is the use of histograms, if only one parameter has to be visualised. Histograms are very useful when the same antigen is to be analysed through multiple experimental samples, such as when studying its frequency or the grade of positivity (in terms of fluorescence intensity – that must be calculated and expressed in a linear scale). Almost all softwares provide tools for the overlay of histograms from multiple samples, allowing a rapid analysis of the parameter(s) of interest.

Differently, when there is the need to visualise two parameters simultaneously, multiple choices are available. Historically, *dot plots* are the most common graph used for this purpose (**Fig. 5.3**). In this type of representation, each singe event is depicted as a *black dot* on the screen. However, monochrome dot plots have many limitations, the first of which is due to the fact that cells displaying the same amounts of fluorescences occupy the same area on the plot and cannot be further distinguished. In fact, when the acquisition of a high number of cells is required, even low cell density-areas are saturated by a consistent number of events, leading to a difficult identification of cellular subsets *(10)*. These hurdles can be overcome by the use of contour plots. In this case, contours identify regions with variable cell density and the representation obtained is irrespective of the number of acquired events (**Fig. 5.3**).

3.6.3. New Tools for the Analysis of PFC Data

The procedures described above are useful when analysing specific subsets and can be utilised when performing multiple, sequential gates, as shown in **Fig. 5.2**. However, in PFC, when several antigens are detected at the same time in a single cell, classical methods of analysis are difficult to employ as they do not consider

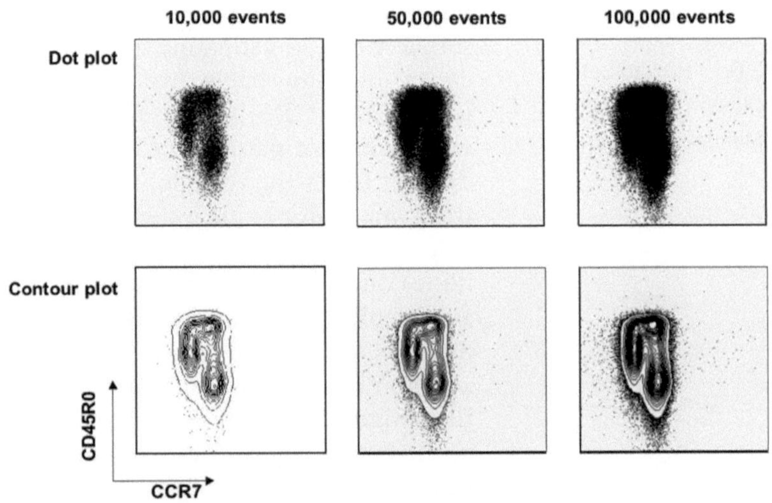

Fig. 5.3. Cellular subsets are better defined by contour plots rather than dot plots. Different numbers of events related to the expression of CD45R0 and CCR7 are visualised in two different ways, as dot plots (*top*) or 5% probability contour plots plus outliers. Note that, in the contour plot representation, the increasing number of events still allows an easy recognition of cellular subsets.

the flow cytometric output as a whole. In other words, the use of 8 mAbs to stain typically 100 μL of blood gives origin to 256 (=2^8) cell subpopulations, not to mention the possibility to add other parameters such as the physical scatters (reaching the number of 1,024 subsets). Methods have been developed to identify subsets that differ among samples, i.e. Probability Binning Comparison *(11)* or algorithms for the automatic identification of subsets *(1)*. Here we describe the procedure to perform Cluster Analysis, an approach we recently used to identify the dynamics of T cell subsets during the ageing of the immune system, and to cluster subjects with similar immunophenotypes *(12)*.

1. Identify the main subset to be analysed (i.e. CD4+ or CD8+ T cells, CD19 + B cells, CD56+ NK cells, etc.);
2. Inside this gate, plot all parameters vs FSC or SSC , or use histograms (**Fig. 5.4a**);

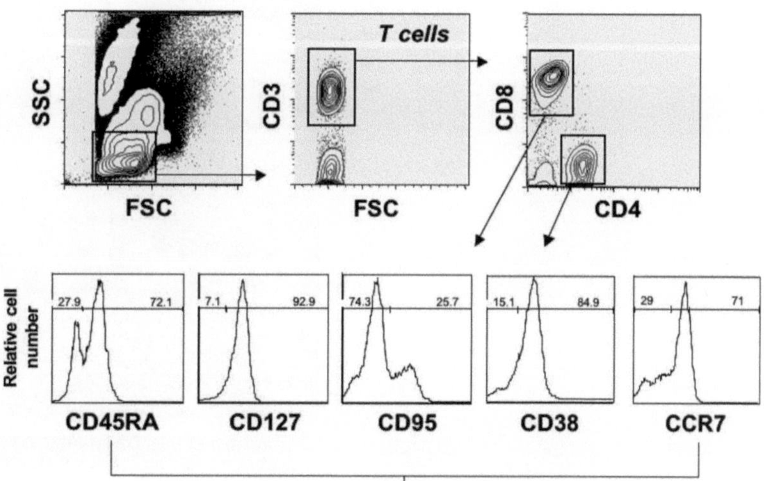

Fig. 5.4. Subjects of different age can be clustered on the basis of the flow cytometric phenotype.
(**a**) PBMCs from a representative donor were stained with eight different mAbs to analyse the expression of CD3, CD4, CD8, CD45RA, CCR7, CD127, CD95 and CD38. Positive and negative expression of antigens is indicated in the histograms. (**b**) All possible phenotypes were generated as described in **Section 3.6.3.** T cell subpopulations were clustered by using "complete linkage" and "correlation similarity" as parameters. Cluster analysis identifies two branches: the one on the left mainly contained young donors, that on the right mainly centenarians. Note that middle-age donors are scattered between the two groups, indicating a high heterogeneity of this group. The grey scale of variables (cell populations) ranges from white (for log ratios of $\leq$3.0) to black (for log ratios of $\geq$3.0).

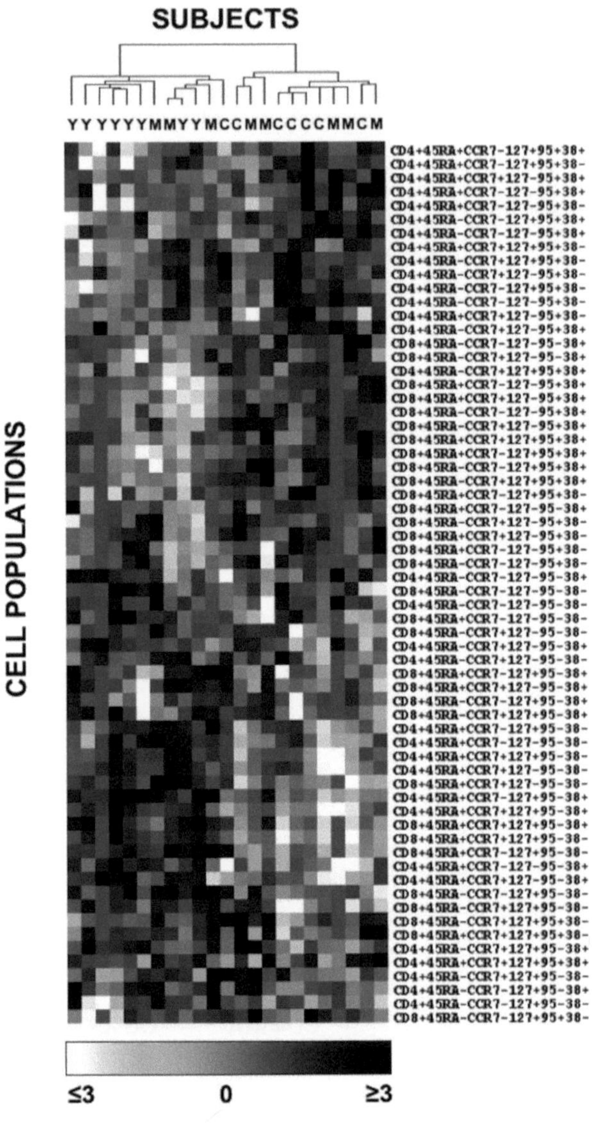

Fig. 5.4. (continued)

3. Identify the threshold of positivity for each antigen analysed, in order to generate two subsets, positive and negative (for example, for CD45R0 antigen identify CD45R0+ and CD45R0- cells). In some cases, positive expression can be further distinguished between *dim* and *bright*, in order to identify cellular subsets with different biological function (for example, CD25 *dim* expression identifies conventional activated CD4+ T cells, while CD25 *bright* expression identifies regulatory T cells);
4. By using a software like FlowJo (Treestar, Ashland, OR, USA), combine positive and negative gates in order to generate all the

possible phenotypes/combinations of antigens (typically 2^n, where n is the number of analysed parameters);

5. Generate a .txt table file which contains all the possible subsets for each experimental sample;
6. Load the .txt table file in Cluster software;
7. Adjust data as following:
   a. Log transform data: replace all data values X by $\log_2(X)$. The advantage of transforming the data into log scale is to allow the identification of the magnitude of change between two or more values, irrespectively of the absolute values in the matrix, i.e. the percentage of the considered subset;
   b. Mean/median center rows and columns: adjust the values of each variable (cell subset/combination of antigens) to reflect their variation from some property of the series of observed values such as the mean or median (identify the distance of each single value to the same reference);
   c. Normalise rows and columns: set the magnitude of a row/column vector to 1.0
   d. Filter data: remove values that do not have certain desired properties from you dataset. Note that, when analysing multiple markers at the same time and when performing multiple gates to identify all possible phenotypes, it is likely that some subsets are present at very low frequency, especially when low number of events are acquired. If needed, these subsets can be removed from analysis by filtering the data;
   e. Hierarchical clustering: cluster rows and columns by choosing the similarity metric and the clustering algorithm;
   f. On Treeview software, visualise the .cdt file generated by Cluster and analyse the data;

For a more detailed description of Cluster and Treeview softwares, download the manual at rana.lbl.gov/manuals/ClusterTreeView.pdf. **Fig. 5.4b** shows the clustering result of three cohorts of subjects of different age (20, 60 and 100 years) on the basis of the 64 possible cell phenotypes of CD4+ and CD8+ T cells identified by combining the expression of CD45RA, CCR7, CD127, CD95 and CD38 expression *(12)*. Note that people with different age can be clustered on the basis of T cell flow cytometric profile. Young individuals go to the left, centenarians to the right, while middle-age donors are scattered between the former two groups.

3.7. General Steps for Optimal Results

1. Establish priorities when designing your multicolour experiments. If working with frozen samples and if looking to rare subsets, always save a channel for a dye able to discriminate dead

cells; then assign the weakest fluorochromes to lineage markers and the brightest ones to antigens with dull expression.

2. Include markers with unknown expression patterns in channels, where the spillover from other fluorochromes is minimal, in order to avoid loss of sensitivity after fluorescence compensation.
3. Start by staining few antigens (3 or 4), then add 1 or 2 antibodies at a time and check whether the initial staining pattern has been modified by the added antibodies. If so, check whether a different staining combination is possible.
4. Always titrate antibodies and dyes before use and choose the concentration at which the best signal-to-noise ratio is obtained. If performing analysis in fixed/permeabilised samples, antibodies should be titrated by using the same protocol and buffers chosen for the experiment. Different protocols and buffers can have diverse effects on dyes and conjugates and, thus, impact on fluorescent signals and compensation.

Acknowledgements

We thank Prof. Wolfgang Göhde (University of Münster), Dr. Luca Cicchetti (Space Import Export, Milan, Italy) and GeneMoRe Italy srl for continuous support. This work was partially supported by grants from Istituto Superiore di Sanità (Rome, Italy), Progetto Nazionale AIDS 2006. Dr. Gabriele Marcotullio is kindly acknowledged for excellent editorial assistance.

References

1. Perfetto SP, Chattopadhyay PK, Roederer M. Seventeen-colour flow cytometry: unravelling the immune system. Nat Rev Immunol 2004;4:648–55.
2. Chattopadhyay PK, Price DA, Harper TF, et al. Quantum dot semiconductor nanocrystals for immunophenotyping by polychromatic flow cytometry. Nat Med 2006;12:972–7.
3. Perfetto S, Ambrozak D, Nguyen R, Chattopadhyay PK, Roederer M. Quality assurance for polychromatic flow cytometry. Nat. Protocols 2006;1:1522–30.
4. Betts MR, Nason MC, West, SM, et al. HIV nonprogressors preferentially maintain highly functional HIV-specific CD8+ T cells. Blood 2006;107:4781–9.
5. Baumgarth N, Roederer, M. A practical approach to multicolor flow cytometry for immunophenotyping. J Immunol Methods 2000;243:77–97.
6. Roederer M. Spectral compensation for flow cytometry: visualization artifacts, limitations, and caveats. Cytometry 2001;45:194–205.
7. Herzenberg LA, De Rosa SC. Monoclonal antibodies and the FACS: complementary tools for immunobiology and medicine. Immunol Today 2000;21:383–90.
8. Sallusto F, Lenig D, Forster R, Lipp M, Lanzavecchia A. Two subsets of memory T lymphocytes with distinct homing potentials and effector functions. Nature 1999;401:708–12.

9. Kantor AB, Roederer M. FACS analysis of leukocytes. In: Herzenberg LA, ed. Handbook of experimental immunology. Oxford, UK: Blackwell Science, 1997:49.1–49.13.
10. Herzenberg LA, Tung J, Moore WA, Parks DR. Interpreting flow cytometry data: a guide for the perplexed. Nat Immunol 2006;7:681–5.
11. Roederer M, Moore W, Treister A, Hardy RR, Herzenberg LA. Probability binning comparison: a metric for quantitating multivariate distribution differences. Cytometry 2001;45:47–55.
12. Lugli E, Pinti M, Nasi M, et al. Subject classification obtained by cluster analysis and principal component analysis applied to flow cytometric data. Cytometry Part A 2007;71:334–44.

Chapter 6

Generation of Human T Cell Clones

Sabrina Mariotti and Roberto Nisini

Abstract

Peripheral blood T lymphocytes are a pool of cells with extremely different characteristics and, therefore, it may be difficult to obtain clear-cut results and to attribute a certain function to a defined T cell population in several experimental settings. The availability of a population of human T lymphocytes deriving from the same progenitor with a unique phenotype and function (clone) may therefore be of help.

This chapter describes the basic procedures that can be used to design a T cell cloning experiment. A prototypic limiting dilution cloning procedure to obtain and maintain in culture peptide-specific CD4 positive T cell clones from peripheral blood is detailed. Methods to generate other types of T cell clones are described in **Section 3.5**. Since peripheral blood is not always the ideal source of lymphocytes to isolate T cell clones, in **Section 3.5** the isolation of cells from biopsies or other tissues is briefly described.

Finally, in the notes, some expedients and wariness that can be useful to improve the efficiency of T cell cloning are reported.

Key words: Limiting dilution, T cell clone, antigen specificity, plating efficiency.

1. Introduction

T lymphocytes are a pool of cells with extremely different characteristics. First of all, every cell that derives from a certain precursor has rearranged its genes for the T cell receptor (TCR) differently from others cells. Thus, there is an extreme heterogeneity of T cells in peripheral blood due to the different TCR expressed by the cells that recognize different epitopes. In addition, T cells can be subdivided in discrete subsets, which are distinguished by the differential expression of molecules, such as CD4 and CD8, but also CD45R, CD25, CTLA-4, chemokyne receptors, etc. These molecules are variably associated to the function that a single T cell is programmed to carry out or to the

Gennaro De Libero (ed.), *T Cell Protocols: Second Edition, vol. 514*

DOI 10.1007/978-1-60327-527-9_6 Springerprotocols.com

differentiation stage in which at a given time that cell belongs. Moreover, the pool of peripheral cells also includes T lymphocytes with different capacity to secrete function-associated cytokines.

Many studies on T cell function may be performed by analyzing peripheral blood mononuclear cells (PBMC), but since PBMC are such a heterogeneous population of cells, in many experimental settings it may be difficult to obtain clear-cut results and to attribute a certain function to a defined T cell population. The availability of a population of T lymphocytes that derives from the same progenitor (clone) with a unique phenotype and function may be of help. On the other hand, limitation on the interpretation of results must be identified also when a clone is used. A clone will consist of cells, which share the same phenotypic characteristics and function, but they may not be representative of the whole population of T lymphocytes in in vivo situations.

T cells recognize their antigens of proteic or lipidic nature presented by antigen presenting cells (APC) in the context of diverse antigen presenting molecules (APM). Upon encounter with a APC presenting a peptide or glycolipid, for example, from bacterial origin during an infection, T cells with a TCR capable of recognizing the complex APM-Ag are activated and may proliferate, giving rise to a progeny of T cells expressing the same TCR *(1, 2)*. Upon activation T cells synthesize their principal grow factor, the interleukin (IL)-2, and express the IL-2 receptor, which enable them to sense the IL-2 presence and continue cell division. The functional consequence of such an expansion is the availability of a large number of T cells specific for the bacterial antigen capable of giving help to B cells or to macrophages, which eventually contribute to the bacterial clearance. With the decay of antigen concentration, due to elimination of the source, i.e., bacterial clearance, the expanded T cell population ceases IL-2 synthesis and reduces the membrane expression of IL-2 receptor, so that cells progressively die. Some cells, however, remain as memory T cells. If a second infection with the same pathogen occurs, APC may present the same antigens and memory T cells are "restimulated" and rapidly expanded with a new cycle of IL-2 synthesis and IL-2 receptor expression. To isolate an antigen-specific T cell clone a strategy is used to in vitro mimic such a physiologic in vivo T cell response.

Five steps may be followed to design a T cell cloning experiment. In the first step, the precise characteristics of the required clones must be identified as well as the criteria to distinguish them from other not relevant clones. Then, the most suitable blood donor to enroll must be defined to succeed in obtaining the wanted clones. The third step includes the methods to enrich the starting pool of T lymphocytes in the required T cell population. The fourth step encompasses the methods to obtain single cell cultures and the last step is the identification, among the growing clones, of those that fulfill the criteria of selection.

1. definition of which clone is required
2. identification of the donor and of the source of lymphocytes
3. strategy to enrich the population to clone
4. methods to obtain a single cell culture
5. selection of relevant clones

1.1. Definition of the Clone Requirements

In the simplest case, it may be needed a random cloned population of T lymphocytes. This is the case of those experimental situations in which it is not relevant the antigen specificity of the cloned population or its function. For example, a cloned population may be an excellent tool to study signal transduction pathways *(3)* or the kinetics of retroviral integration *(4)*, or even for crystallographic studies of the TCR *(5)*. In this category of non-selected T cell clones requirement also fall those studies that are aimed at verifying the frequency of a given character in an array of casually generated clones *(6)* to calculate the frequency of occurrence of the characteristics *(7)*. In the majority of cases, however, T cell clones are required for having certain characteristics. It may be wanted an array of T cell clones with defined antigen specificity, or with helper or cytotoxic function, or with a certain phenotype, i.e., $CCR3^+$, $CD30^+$ or $V\beta 11^+$. The precise definition of the required populations and the strategies to identify such population are essential in planning the next steps of T cell clone generation and finally in obtaining a positive result.

1.2. Identification of the Donor

Obviously, in the case of non-antigen-specific T cell clones there is no need to choose a specific donor as source of lymphocytes, thus the procedure may proceed directly to step 4. In those cases, instead, where the interest is focused on generation of T cell clones with certain antigen specificity, the source of lymphocytes becomes an important issue. For example, to obtain a single clone of antigen-specific T lymphocytes requires methods to identify and select the lymphocytes with the desired specificity among the pool of cells with different specificities present in the peripheral blood. Thus, the chances to succeed will be proportional to the frequency of the antigen-specific cells among the whole population of T lymphocytes in the blood, and this implies that the required populations of T cells must be present in the peripheral blood of certain donor. For instance, it will be relatively easy to isolate tetanus toxoid (TT) specific T lymphocytes from a healthy donor recently vaccinated against tetanus, while it will be less easy to isolate the same clones from a donor vaccinated several years before and it will be uncertain to obtain clones with the same specificity from a donor who has never been vaccinated against

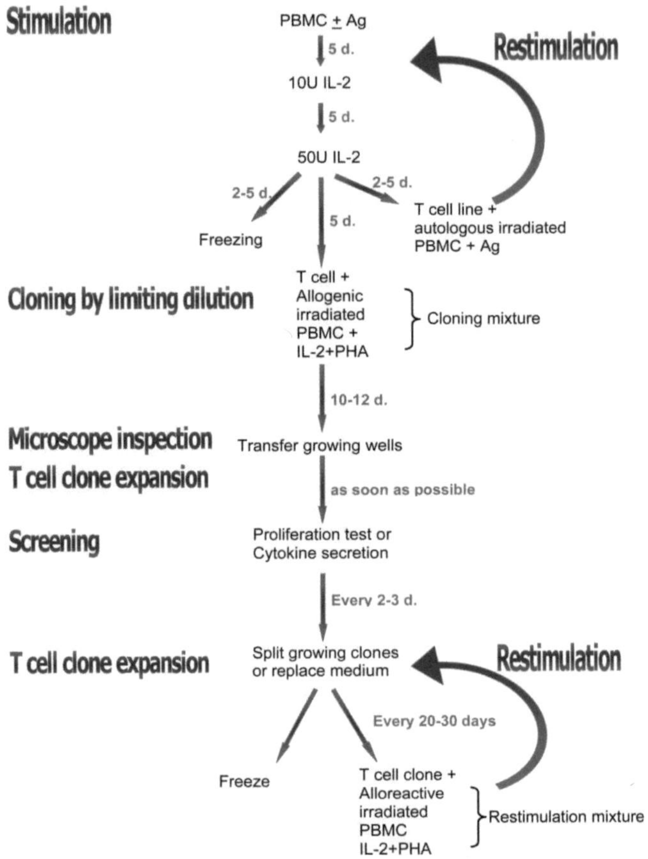

Fig. 6.1. A scheme of enrichment of a low frequent antigen-specific T cell population among the whole heterogeneous population of T lymphocytes.

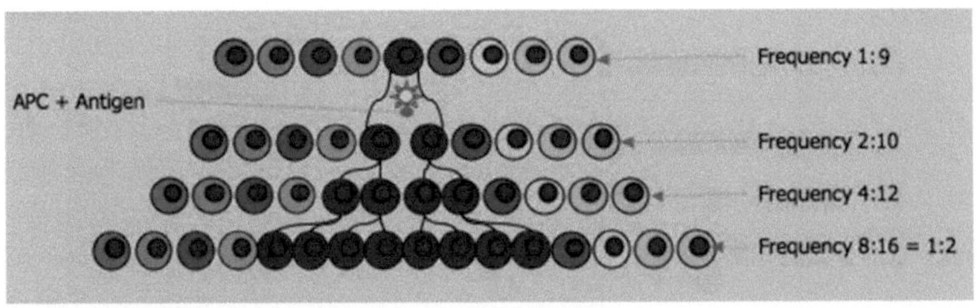

Fig. 6.2. General diagram for T cell cloning and maintenance in continuous culture.

tetanus. Thus, recently vaccinated subjects represent a source of antigen-specific lymphocytes to enroll for the isolation of T cell clones *(8)*. Other sources for T cell clones isolation are represented by patients with or recently recovered from a certain disease, which is associated with an antigen-specific T cell expansion. Examples of such diseases are infectious or allergic diseases *(9, 10)* and, with several limitations, also neoplastic *(11)* or autoimmune diseases *(12)*. In most of these cases, peripheral blood is the ideal source of lymphocyte to isolate T cell clones. There are, however, situations in which alternative sources of lymphocytes are preferred. This is the case, for example, of lymphocytes infiltrating tumors or inflammatory sites. In this case, lymphocytes could be isolated from biopsies *(13)* or other tissues after surgery *(14, 15)*. Isolation of lymphocytes in these situations will be briefly described in **Section 3.5**. T cell population in the blood is heterogeneous, thus, if the required T cell clones are frequent, there is no need to use methods for its enrichment: for example, we would require non-specific CD4 or non-specific CD8 T cell clones. In this case, using a limiting dilution method to obtain clones, we will have almost half of the clones with the desired characteristic.

1.3. Strategy to Enrich the Population

On the other hand, if we need to clone rare T cells, the strategy to enrich the population of required cells becomes an important issue. For example, before cloning a low frequent T cell population, such as a population bearing a certain Vβ chain as part of its TCR, we might need to enrich it, by magnetic- or flow cytometric-sorting. In the case of antigen-specific T cells, that cannot be identified by their phenotype and whose frequency is very low, a primary culture to enrich in antigen-specific cells is required (**Fig. 6.1**). This situation will be used as an example of the experimental procedures to generate a cloned population of T cells, *to obtain single cell cultures* and to *select the relevant clones*, because it represent the most stringent procedure for experimental requirements. It will be described a prototypic cloning procedure to obtain and maintain in culture peptide-specific CD4 positive T cell clones (**Fig. 6.2**) and the differences to generate other types of antigen-specific T cell clones will be briefly described in **Section 3.5** *(16)*.

2. Materials

Most of the materials can be indifferently purchased from any company. The companies are specified only when necessary. Here is listed all the material used in the examples, but not all the listed items are required in dependence of the wanted T cell clone population.

1. Serum free complete culture medium: RPMI 1640 (without glutamine) supplemented with 100 U/ml kanamycin (or others comparable antibiotics), 1 mM glutamine, 1 mM sodium pyruvate, 1% non-essential amino acids

2. Sera: 5% autologous serum (*see* **Section 3.5**) or 5% human serum or 10% fetal calf serum (FCS) heat inactivated (30 min, 56°C)
3. Human recombinant IL-2 (Chiron Corporation; Hoffmann-LaRoche)
4. Heparinized venous blood, 20 ml as source of PBMC for primary culture, screening of clones, and BCLL production.
5. Non heparinized venous blood, 10 ml for autologous serum collection (*see* **Section 3.5**)
6. PBMC purified from heparinized blood by Ficoll–Hypaque gradient
7. Purified antigen (*see* **Note 1**)
8. Humidified 37°C, 5% CO_2 incubator
9. Trypan blu
10. Counting slides (Boyden chambers or analogues)
11. Purified phytohaemagglutinin (PHA; indirect TCR cross-linking) (HA-16 from Murex Diagnostics)
12. Cobalt source for irradiation (*see* **Section 3.5**)
13. Mitomycin C (DNA cross-linking reagent), it is light sensitive (*see* **Section 3.5**)
14. Cloning mixture: complete medium, 5% human serum (10% FCS may be used but the efficiency of the cloning may be reduced), 100 U/ml IL-2, allogeneic irradiated (or mitomycin C treated, *see* **Section 3.5**) PBMC (at 4×10^5 cells/ml), PHA (1 μg/ml)
15. Restimulation mixture for clones: complete medium, 5% human serum or 10% FCS, 100 U/ml IL-2, allogeneic irradiated (or mitomycin C treated, *see* **Section 3.5**) PBMC (at 1×10^6 cells/ml), PHA (2 μg/ml)
16. Tritiated ^{3}H-thymidine (Amersham; DuPont NEN; ICN Biomedicals). It is a radioactive material that require a special handling and discard procedures (to be used if proliferation assays are required)
17. Automated multiwell harvester (*see* Note to step 16)
18. Scintillation counter for beta particles (*see* Note to step 16)
19. Cell culture plastic plates: 96 and 24 wells flat-bottomed and 96 wells U-bottomed plastic plates
20. Plastic plates for cell cloning: Terasaki plates (Nunc) or 96 wells U-bottomed plastic plates
21. Multistep device or multisyringe device for Terasaki plates cloning (Hamilton)
22. Multipipette with 8 or 12 channels (in the case of cloning in 96-wells plates)

23. Human IFN-γ ELISA (to be used if cytokine secretion assays are required)
24. ^{51}Cr. It is a radioactive material that require a special handling and discard procedures (to be used if cytotoxicity assays are required)
25. Gamma-counter (*see* Note to step 24)
26. Freezing medium: heat inactivated FCS plus 10% dimethyl sulfoxide (DMSO)
27. B95.8 cell line (marmoset cell line; American Type Culture Collection ATCC CRL 1612) for generation of EBV-containing supernatant (*see* **Section 3.5**). Biosafety practices must be followed
28. Cyclosporine A (Sigma; Calbiochem; Novartis Pharma B.V.)
29. EBV transformed B cell Line BCLL (*see* **Section 3.5**). Biosafety practices must be followed.
30. Nylon cell strainer 40 μm (Falcon)

Important: Blood, cells, and serum of human origin are potential sources of infectious and transmissible diseases and human blood, serum and cells as well as all the cultures containing human serum and cells must be handled and discarded according to biohazard procedures. All the procedures for culturing human cells must be carried out in biohazard level hoods.

3. Methods

A T cell cloning is in theory a simple procedure: if we presume that the blood of a given donor contains the desired T cell population to obtain as a clone, we could isolate PBMC, count the cells and dilute the cells up to a concentration which would allow to seed only one cell in a single well and induce the proliferation of single cells with appropriate stimuli, such as PHA. We will obtain a high number of growing cells that could be screened in order to find the clones that satisfy our requirements. In practice, this procedure is almost impossible to succeed or it could be useful with an absolute negative cost–benefit ratio and a high dose of luck. Since T cells specific for a given antigen are relatively rare among the whole population of T cells that is represented by heterogeneous cells with specificity for virtually all antigens, the above mentioned procedure would require to seed thousands of wells. Moreover, a single dilution of the cell population does not assure that cells can be effectively separated in one for well as expected. A procedure of enrichment of antigen-specific cells and a method to reduce the experimental variability of the limiting dilution procedure is thus required to increase the cost–benefit of the cloning method.

3.1. Primary T Cell Culture

PBMC are resuspended in complete medium + 5% autologous serum at the concentration of 1×10^6 cell/ml. Cell suspension of 100 μl are distributed in flat-bottomed 96-well plates. In a typical experiment, 80 wells will be filled. Sixteen and 64 wells will be filled with 100 μl of complete medium + 5% autologous serum without and with antigen, respectively (**Fig. 6.3**). Cultures without antigen will indicate the control basal growth of cells without antigen and will monitor whether the culture conditions are suitable for the isolation of antigen-specific T cells. In these wells, if the culture medium does not contain mitogens or unwanted antigens (present, for example, in the serum), T cells should not grow and would represent the "negative cultures". It is important to envisage in the experimental design such a control culture, since an unwanted growth in these wells will indicate that a T cell division occurs independently on the antigen and will prevent to proceed in the next steps. A typical example of the helpfulness of control wells can be given by experiments made in FCS instead of autologous or human serum: FCS contains antigens of bovine origin as well as low endotoxin level and human T cell mitogens that induce the growth of a significant number of cells. The number of FCS responder cells will be higher than the number of T cell specific for a given antigen, rendering difficult if not impossible to isolate the desired T cell clones. Usually antigen can be added at concentration ranging from 1 to 20 μg/ml (*see* **Note 1**). However, it is not possible to provide a general rule. Since the real antigen concentration is not always what measured with colorimetric methods and since the frequency of antigen-specific cells varies as well the efficiency of APC in the wells, unless tested in advance, it is recommended to use more than one concentration of antigen (*see* **Note 2**). Thus, cultures of at least 12 replicates with different antigen concentrations may be required to find the dose that results in the best proliferation of specific cells.

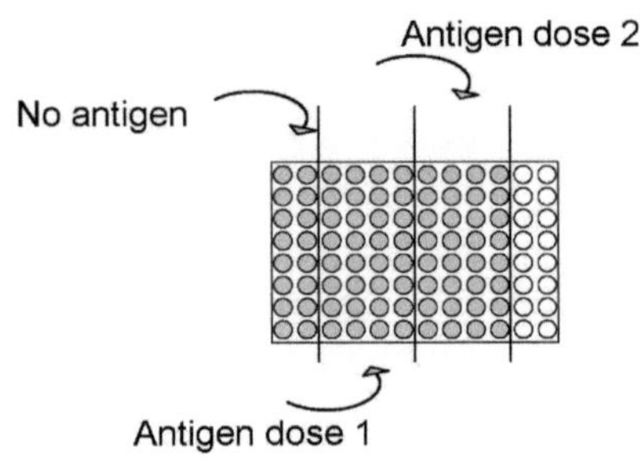

Fig. 6.3. A typical primary culture experimental design using 96 well flat-bottomed plastic plates.

In a prototype experiment to isolate TT specific T lymphocytes from a recently vaccinated subject, 32 wells with a final concentration of 1 μg/ml and 32 wells with final concentration of 10 μg/ml TT are suggested. Antigen is diluted at a 2 × concentration in complete medium + 5% autologous serum and then 100 μl of the diluted antigen are added to the 100 μl of complete medium + 5% autologous serum containing PBMC already distributed into the wells. Thus, each well will contain 200 μl of complete medium + 5% autologous serum, 1×10^5 PBMC with or without antigen at two concentrations (**Fig. 6.3**).

Plates will be kept in incubator for 5 days and then microscopy inspected.

A good culture will end up with no T cell growth in negative control wells and with clumps of cells distributed onto the well bottom surface with a frequency dependent on the antigen dose. After inspection, at day 5, 10 μl/well of 10 U/ml IL-2 in complete medium + 5% autologous serum medium will be added. A low IL-2 dose will now facilitate the proliferation of those cells that, following antigen recognition on the surface of APC have the IL-2 receptor CD25 up-regulated.

At day 10, a new microscopy inspection will show that some IL-2 mediated T cell growth occurs in control wells without antigen. This proliferation, however, if the culture conditions are correct, will be limited and sporadic. On the other hand, in wells with antigen, the proliferation will be evident and some cells will now be found out of the clumps with blast morphology. At the same day 10, 10 μl/well of 50 U/ml IL-2 in complete medium + 5% autologous serum medium will be added.

Daily inspection will indicate the preferred day to proceed with the cloning experiment. The preferred day will be that day, usually the 12–15th, in which cells in the well with antigen are growing making the medium to turn pink–yellow, indicating the acidification of the medium and suggesting the need to split the medium, while cells in the wells without antigen are not yet beginning to outgrow. If cells in the wells with antigen are growing and the medium turned to yellow, but the cloning experiment cannot be performed and must be delayed, wells can be split into two. In this case a twin plate can be prepared. Using a 8- or 12-channel multipipette, mix accurately each well, and transfer 100 μl in the second plate mirroring the scheme of the first, changing the tip for every row (**Fig. 6.4**). Then, add to each split well 100 μl of complete medium + 5% autologous serum medium containing 50 U/ml of IL-2. Wells containing cells without antigen must be treated with the same procedure. If the cloning experiment will be delayed for more than a week, pools of growing cells from different wells containing the same antigen concentration can be made and cells can be counted and frozen (*see* **Section 3.5.6**). It is in fact not suggested to maintain in culture these lines: non-antigen-

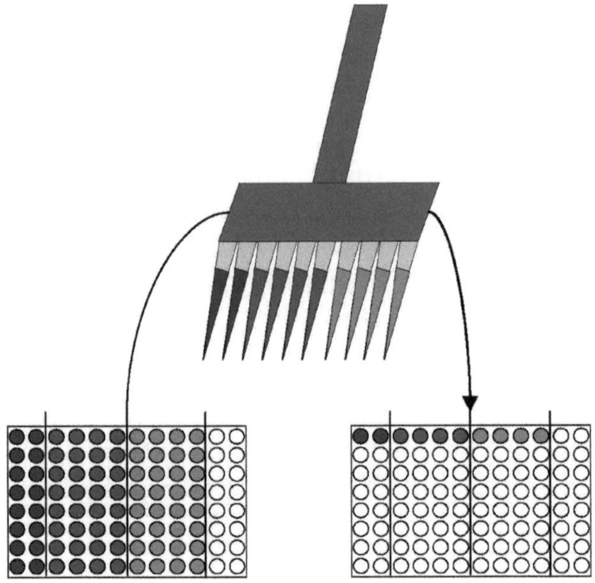

Fig. 6.4. Lines growing in 96 well flat-bottomed plastic plates can be split using a 8- or 12-channel multipipette, transferring to a twin labeled plate half of each well content after pipetting and changing the tips after the transfer of every row. When all the row are transferred, with the same pipette fresh culture medium is added to the two plates, to re-establish the same volume of culture medium.

specific cells can in fact outgrow and reduce the frequency of the antigen-specific cells, reducing the efficiency of antigen-specific T cell recovery after cloning. An alternative to clone the lines at this point of the culture is represented by the possibility to restimulate them with APC and antigen (*see* **Section 3.4** *and* **Note 3**). Moreover, if the required T cell population has a very low frequency in PBMC, it could happen that cells in the wells with antigen are not expanded enough at day 12–15 of primary culture. In this situation, it is opportune to restimulate the T cell line with autologous APC plus antigen to allow a further expansion of antigen-specific T lymphocytes (*see* **Section 3.4** *and* **Note 3**).

After 12–15 days of primary culture in optimal conditions, T cells specific for the antigen had been stimulated and had begun to proliferate, so that their frequency at this stage is increased. However, it is well established that during the same period of culture, several other cells with different specificities are non-specifically activated and do proliferate indeed *(17)*. Thus, the 12–15 days culture contains blasts cells specific for the desired antigen, but also blast cells and resting cells with non-relevant specificity. If antigen-specific cells were present with a theoretical frequency ranging from 1:10,000 to 1:100,000 before the primary culture, after the 12–15 days of culture in the appropriate conditions their frequency may increase up to 1:10–1:100. This means that we have chances to find an antigen-specific cell for every 10–100 cells if we could separate each cell from others and allow it to grow as a clone, i.e., as the progeny of a single cell. The remaining 9–99 cells represent cells with different specificity that underwent non-specific activation or resting cells that did not die during the culture. The cloning procedure will allow separating these functionally different cells and screen for the antigen-specific cells we want to obtain.

3.2. Cloning by Limiting Dilution

Lines at day 12–15 of primary culture or at day 12–15 after restimulation (*see* **Section 3.4**) or thawed after freezing (*see* **Section 3.5**) can be cloned with different procedures, but only the limiting dilution method will be described here.

3.2.1. Limiting Dilution Principle

The most used method to allow the growth of the progeny of a single cell is the limiting dilution technique *(18, 19)*. The principle of this method relies on the possibility to dilute the cultures so that in a given volume is theoretical present a single cell. For example, if the 12–15 days culture contains 2×10^5 cells in the volume of 200 μl, it means that this culture contains 1,000 cell in 1 μl. If we dilute the culture 1:1,000 by picking up 1 μl of the culture and adding 999 μl of medium, the obtained diluted culture will contain 1,000 cells in 1,000 μl. Thus, every μl of this diluted solution will contain one cell that can be placed into a single well and cultured in the appropriate condition to allow its grow as a clone.

In practice, there are several problems that limit the efficiency of the procedure as described. For instance, the cell count by Boyden chambers or even by electronic counters represents a typical example of count by sample and not of the cell population to be used. In addition, the count by sample is characterized by the experimental variability that makes it almost impossible to have the real enumeration of cells, so that, practically, we may have just a good estimation of the real cell concentration. Moreover, due to the fact that often T cells grow in clumps where many cells are attached one to the others, and the electrostatic forces and superficial tension at the interfaces can prevent an equal distribution of cells in the diluted volume.

The limiting dilution method that is described below for the isolation of T cell clones take into considerations the problems and constraints described above, but further check for clonality of the obtained populations must, however, been programmed *(20)*.

3.2.2. Selection of T Cell Pool to be Cloned

After the 12–15 days of primary culture, the growth of cells is not always homogeneous in the different wells. A good procedure is to select and pool those wells in which cells are growing the most. Coming back to the first example, among the 32 wells with 1 μg/ml of antigen we could expect to obtain eight wells with a good T cell growth, while among the 32 wells with 10 μg/ml of antigen we could obtain 16 wells in which the growth of cells is microscopically much higher than in control wells without antigen. Since each well contains 200+10+10 μl of medium, we could then make a pool of about 1.7 ml from the first eight cultures and another pool of 3.5 ml from the other 16 cultures in two plastic tubes. We can then make a third pool, mixing part of the first and the second pool, that will be used for the limiting dilution and cloning. The cells in the first two pools can be frozen or maintained in culture or restimulated, as specified in **Section 3.4**.

The third pool of cells will thus contain cells grown in the presence of both low and high dose of antigen and a certain number of these cells have good chances to be antigen-specific. Since the cells in the pool derive from cells independently grown in different wells, we can reasonably hypothesize that they represent a variety of antigen-specific cells and not the progeny of a single growing cell *(21)*.

3.2.3. Preparation of T Cell Dilutions

To proceed to the calculations for the limiting dilution, we need to know which will be the final volume of medium that we will use for the single cultures. If we are going to use Terasaki plates, every well of these plates can be filled with a final volume of 20 μl. If we are going to use 96 well U-bottomed plastic plates, 100 μl will be the final volume of the cultures. The choice of the plate is irrelevant for the scope of the limiting dilution, but every plate has its pros and its cons. Terasaki plates require the minimum amounts of cloning medium, allowing saving components and, in particular, in IL-2 and feeder cells, but multipipette device for these plates are not currently found in all laboratories. Therefore a multistep device is required to dispense the cells in the well and with these pipettes the procedure is time consuming, boring, and a good experience of the operator is required since many times the medium does not rich the well, but is dispensed outside. Moreover, the small volume of medium per well (20 μl) imposes perfect incubation conditions to avoid fast evaporation and therefore failure of the procedure. On the other hand, 96-well plates are easier to manage than the Terasaki, but they require more room in the incubator, higher amounts of medium components, and higher number of feeder cells than Terasaki plates.

As an example, it is described the procedure to perform a typical cloning experiment using Terasaki plates. In Terasaki plates, the final volume of each well will be 20 μl. If we could be precise in the cell count, we could make calculations to dilute cells in order to have 1 cell/20 μl per well. For the reason described above, it is not useful to trust the counting methods. If an error occurs in any step of the counting process, we could easily make a dilution that is 0.1 or 0.01 cell/20 μl, that means that we will have a well containing a growing cell every 10 or 100 wells. If we design an experiment using 1,000 wells, we could obtain only 100 or 10 growing cells, respectively. Since the frequency of antigen-specific T cells will be 1:10–1:100 we could obtain one or none antigen-specific T lymphocyte. On the other hand, if an error occurs and the real concentration of cells upon dilution will be of 2 or 10 cells/well, we could have en extremely high number of wells containing growing cells that will require a complex screening, but the positive cells that we will select with high probability will not be the progeny of a single cell. A strategy that can be used to ride over these problems is to make at least three different dilutions and analyze statistically the number of growing cells for every dilution to identify the dilution that with higher probability results as the more accurate *(9, 18)*. For example, we can count cells and dilute them in order to have a concentration of 5 cells/well, than make a 1:5 dilution that will give rise to a concentration of 1 cell/well and a further 1:5 dilution to obtain a final concentration of 0.2 cells/well. With this strategy, there is a higher probability to find a concentration of cells that will be adequate. In the case of Terasaki plates, to obtain a concentration of 5 cells/well we need a concentration of 5 cells/20 μl that corresponds to 250 cells/ml (**Fig. 6.5**). In the case of 96-well plates, if we want to seed 5 cells/well we have to prepare a dilution of 5 cells/100 μl, i.e., 50 cells/ml *(22)*.

Continuing with the above reported example, we could have obtained a pool of cells deriving from antigen-activated cultures of T lymphocytes of which we should determine the concentration.

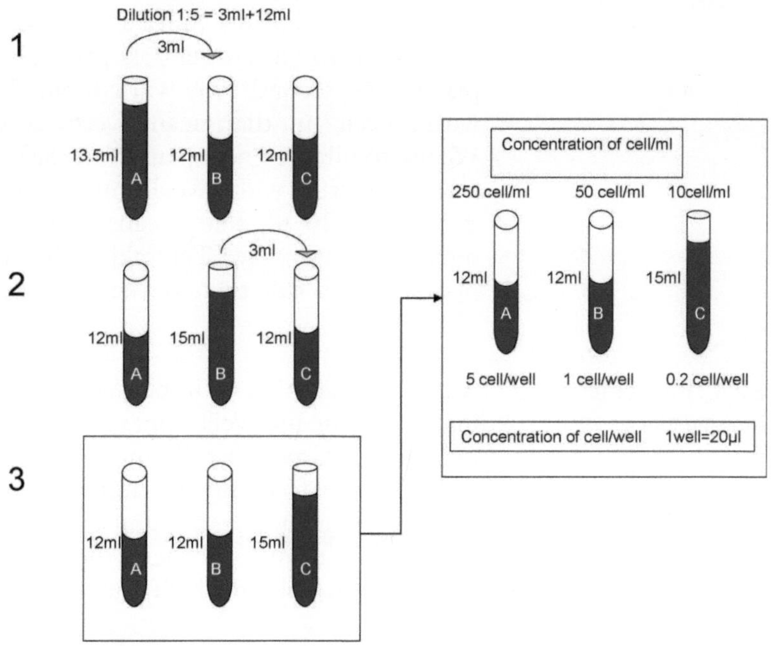

Fig. 6.5. Limiting dilution diagram.

As described above, the counting of the T cell line is a critical step of cloning procedure. It should be performed using a vital stain like Tripan Blue that allows to exclude dead cells and mixing up very well the suspension of cells trying to disrupt clumps. If the concentration of T lymphocyte pool is, for example, of 4×10^5 cells/ml, to obtain a concentration of 250 cells/ml (that will give a dilution of 5 cells/well) we have to make a 1:1,600 dilution (400,000/250=1,600). The dilution 1:1,600 can be dividend in three serial dilutions: 1:16, 1:10, and a last 1:10. We can mix up very well the pool of cells trying to disrupt clumps and cell aggregates by passing in serial tubes with micropipettes the same volume of, for example, 500 μl, changing the tip at every passage. Then, we can make the first dilution (1:16) by adding 100 μl of our pool to 1,500 μl of medium in a new plastic tube. The tube is then vortexed and the second dilution (1:10) can be performed moving 500 μl of the first dilution to a new tube containing 4.5 ml of medium. We can now prepare three tubes (**Fig. 6.5**), the first containing 13.5 ml, and the other containing 12 ml of cloning mixture (*see* **Section 2**).

The last 1:10 dilution will be performed moving 1.5 ml of the previous dilution to the first tube (A) containing 13.5 ml of cloning mixture. This tube will be vortexed and then 3 ml will be moved to the second tube (B). Lastly, after vortexing, 3 ml will be moved from tube B to tube C, which will be vortexed before use.

According to this dilution, the first tube will contain 12 ml of cloning mixture with 250 cell/ml, that means that it will contain 5 cells in 20 μl, that represents the volume to be added to every well of Terasaki plates. Since a Terasaki plate has 60 wells, tube A is sufficient to fill ten Terasaki plates, however, we use to fill five plates. The second tube will contain 12 ml of cloning mixture with 50 cell/ml, that means 1 cell/20 μl or 1 cell/terasaki well. We use to fill up to nine plates. The last tube will contain 15 ml of cloning mixture with 10 cell/ml, which means that it will contain 0.2 cell in 20 μl, that means that one cell will be randomly distributed every five Terasaki wells. We use to fill 12 Terasaki plates. Plates will then be kept in incubator for 12 days (*see* **Note 4**).

3.2.4. Transfer of Growing Wells

After 12 days, plates will be microscope inspected using a 20 × magnification and wells containing growing cells will be marked (**Fig. 6.6**). Growing cells must now be transferred into larger wells. We use to count the number of cultures to be transferred and prepare an adequate number of U-bottomed 96-well plates, making care to mark properly the new plates to identify cultures derived from wells in which dilution of 5, 1, or 0.2 cell/well were seeded.

The content of each positive Terasaki well will be transferred into 96-well plates U-bottomed prefilled with 180 μl of complete

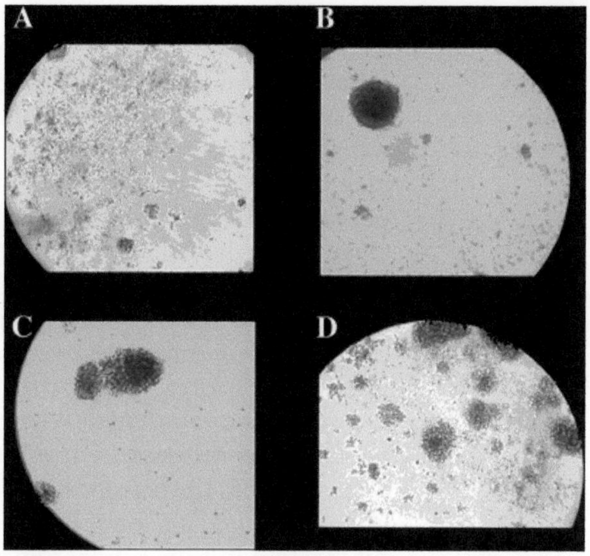

Fig. 6.6. Microphotography of Terasaki plates. (a) and (b): Negative wells in which a clone is not grown. A growing clone after 7 (c) and 12 days (d) of culture.

medium + 5% human serum or 10% FCS and IL-2 at 100 U/ml changing the tip for every clone. We use to prefill one out of three wells in horizontal rows of the 96-well plate, so that clones that grow more than others can eventually be split into the neighbor wells (**Fig. 6.7**). Cells in these plates can be taken in culture for several days until screening or restimulation, making daily inspection to split the cultures when required or replacing half of the medium every 2–3 days.

3.2.5. Plating Efficiency

To calculate the efficiency of the cloning and to estimate whether cells in culture can or cannot be statistically considered as progeny of a single cell, the following procedure must be followed. We have to enumerate the wells filled for every dilution. In the

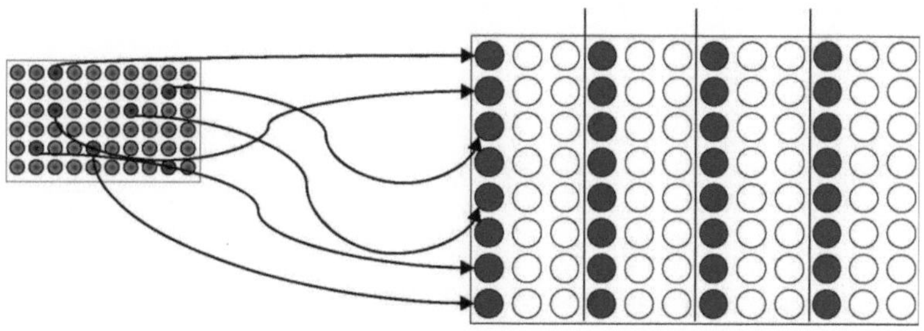

Fig. 6.7. After cloning, growing cultures must be transferred from cloning plates to new plates. Positive wells (in *red*) can be transferred into wells with two adjacent wells left empty. In this way, if some clones grow faster than others, they can be split into the closest wells until all the clones are grown and can be screened or restimulated.

example, we have filled five Terasaki plates with cells at 5 cells/well, nine plates at 1 cell/well, and 12 plates at 0.2 cell/well. It means that we have filled 300 wells at 5 cells/well, 540 wells at 1 cell/well, and 720 wells at 0.2 cell/well. Then, the number of wells without any growth must be enumerated and the percentage of negative wells calculated for each dilution. These numbers allow us to draw the so-called plating efficiency chart (**Fig. 6.8**) that shows the obtained percentages of negative wells in function of the number of cells seeded per well. It is intuitive that the number of negative wells will be low in those plates, where 5 cells/wells were seeded, and that this number will increase in plates filled with cells at lower concentrations. However, the experimental confirmation of this trend is important, since it indicates that dilutions have been performed correctly and that the distribution of cells upon dilution was free and not constrained by, for example, cell adhesiveness or electrostatic forces. In addition, a linear increase of the number of negative wells and the confrontation with the theoretical increase that should be observed as obtained in several experimental settings for statistical purposes *(18, 23)*, will give a tool to judge whether the clones we obtained can be statistically considered as progeny of a single cell. A cell culture, even if homogeneous, represents a population of individual cells programmed to have a certain life span and metabolic fate. Not all the cells of a culture will divide and grow, some of them being, for example, at the end of the life cycle or have begun a programmed cell death due to environmental conditions. It has been demonstrated with several experimental procedures, including the physical seeding of single cells in individual wells using a microscopic micromanipulator, that when single cells of a given population are independently seeded in different wells, as an average only 63% of them divide determining the growth of a progeny. It means that in

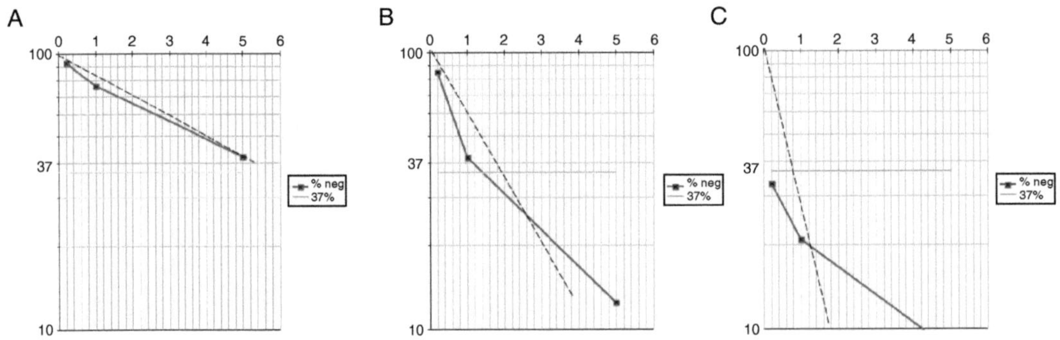

Fig. 6.8. Three examples of plating efficiency (*see* the test). *X*-axis represents the number of plated cells per well; *Y*-axis represents the percentage of negative wells. *Red lines* represent the obtained plating efficiency chart, the *dotted lines* are the best fitting lines and the *green lines* indicate the 37% of negative wells.

such controlled experiments it has been observed that on average 37% of wells will not contain growing cells *(24, 25)*.

Figure 6.8 reports an example of results obtained in three independent experiments in which the plating efficiency varies notably. In the first experiment (A), the percentage of negative wells obtained with all the dilutions always is higher than 37. We can conclude that even if cells were diluted to a theoretical concentration of 5, 1, and 0.2 cell/well, due to errors of counting or to the characteristic of the cell line, in practice less than one cell per well grew in all the seeded wells. Practical consequence of this is that all the lines independently grown that have been obtained are likely to represent progeny of single cells, i.e., clones. On the other hand, in the experiment (C) the T cell lines obtained from the cultures in the wells in which 5, 1, and 0.2 cell/well were seeded, according to the dilution initially made, cannot be considered as clones, because the percentage of negative culture is below 37 in all the cultures. The most likely explanation for this result is that the number of cells in the primary culture pool was underestimated, so that the dilution corresponding to 0.2 cell/well probably contained more than 1 cell/well, and consequently a higher number of cells than the estimated was seeded in the other conditions too. Lastly, experiment (B) reproduces the result of the plating efficiency more commonly obtained in which the experimental frequency is similar to what expected. In this experiment, in fact approximately 48% of negative cultures were observed in the group of plates, where 1 cell/well was seeded.

A practical consequence of the results shown with this analysis is that if antigen-specific responses will generate the line depicted in experiment (C), a more accurate evaluation of the responder line should be made, since there is an high probability that the positive cell line will not be a clone, but a mixture of cells not all antigen responder. On the other hand, the high number of T cell lines obtained in the experiment (A) will increase the chance to find an antigen-specific T cell clone also among those lines derived from plates in which 5 cells/well were theoretically seeded.

3.3. Screening Tests

The clones obtained with the limiting dilution method (in the example, all the clones in **Fig. 6.8a** and the clones derived from wells seeded at 0.2 and 1 cell/well in **Fig. 6.8b**) bona fide represent populations of cells deriving from single progenitor. However, since cells were stimulated with a non-specific mitogen (PHA) the obtained clones derive from all the living cells that were present at the end of the primary culture: resting cells, non-specifically activated cells and antigen-specific cells. The more efficient was the primary culture in increasing the frequency of antigen-specific cells, the higher will be the number of antigen-specific clones that it will be possible to select among all the growing ones.

An antigen presentation test must be performed to select the antigen-specific among all the growing clones. The read-out of the test can be a proliferation assay, since T cells which recognize the antigen undergo cell division and proliferation, or cytokine secretion measurement. Proliferation of a T cell clone can be measured as ^{3}H-thymidine incorporation (*see* **Section 3.5**) and cytokine secretion by ELISA (*see* **Note 5**).

Irrespective of the read-out, every antigen presentation test requires the availability of autologous or compatible APC and adequate amounts of the same purified antigen used for primary culture.

The most useful source of APC is represented by BCLL. If it were possible to produce a BCLL from the PBMC of the donor (*see* **Section 3.5**) at the time in which an antigen presentation test is required, BCLL are usually expanded enough to be used as APC. If such APC were not produced or if they are not expanded enough to be tested, a further isolation of PBMC from the original donor is required, unless an aliquot of PBMC were not frozen at the time of the first isolation. PBMC will then be used as APC.

In a typical ^{3}H-thymidine incorporation based proliferation assay, that until now represents the easier way to measure the antigen responsiveness of a T cell clone, the APC are resuspended in conical tubes in complete medium with serum at a maximal concentration of 1×10^6 cells/ml and divided into two aliquots: the antigen is added in one of the two tubes at the same concentration used for primary culture. APC are kept overnight at 37°C in the incubator to allow internalization and loading of the antigen. Then the APC must be irradiated or mitomycin C treated (*see* **Section 3.5**) to block their proliferation, which would overwhelm the proliferation of T cell clones. After irradiation (or mitomycin treatment) the APC are washed, resuspended at $2–5 \times 10^4$ cells/ml in case of PBMC or 3×10^4 cells/ml for BCLL and 100 μl of cell suspensions are distributed in flat-bottomed 96-well plates. For screening purposes, for every clone to test one well will be filled with APC without antigen and another adjacent well with APC previously pulsed with antigen. T cell clones are collected, extensively washed to completely remove IL-2 and resuspended with serum complete medium to obtain a concentration of approximately 3×10^4 cells/ml. Cells will be added (100 μl) in both the well previously filled with antigen pulsed and non-pulsed APC. Plates will then be left into the incubator (*see* **Section 3.5** *and* **Note 6**). After 2 days of culture ^{3}H-thymidine is added and cell will be harvested for counting after 8–18 h (*see* **Section 3.5**). For each clone a stimulation index (SI) based on the obtained counts per minute (cpm) will be calculate as follows: cpm in the culture with antigen/cpm in the culture without antigen. Clones with cpm >1,000 and SI >3 will be considered antigen-specific.

In alternative, or when radioactive material cannot be used, the antigen-induced T cell activation can be evaluated measuring their cytokine secretion by ELISA. In this case after the 2 days, the culture supernatant can be collected and the T cell-released cytokines quantified. The SI will be calculated accordingly.

To screen the clones, part of each clone was tested using APC with or without antigen and part was kept in culture. The clones that scored positive in the antigen presentation test, must now be identified among the others in the culture plates. At this point of the procedure these clones are "old cells": they were stimulated with PHA for the limiting dilution and remained in culture for at list another week after the 12 days of incubation in Terasaki plates. Their growth-rate is decreased and they are acquiring the morphology of resting cells, accordingly. They need to be restimulated to be expanded and to be used as tools for experiments. Specific clones can be moved into larger wells and treated as follows.

3.4. T Cell Restimulation

Both T cell lines and clones may need to be restimulated.

3.4.1. Restimulation of Antigen-Specific T Cell Lines

After the primary culture, if the cloning cannot be performed or must be delayed for more than a week or, lastly, if lymphocytes are not enough expanded after the primary culture, the T cell line can be restimulated using the same antigen used for primary culture and autologous or compatible APC. For this additional step further isolation of PBMC from the original donor is required, unless an aliquot of PBMC was frozen at the time of first isolation. These PBMC will be irradiated and used as APC cells in co-culture with T cell line in a ratio 1:1. After irradiation PBMC are washed and resuspended at $0.5–1 \times 10^6$ cells/ml and 100 μl or 1 ml placed in 96- or 24-well plates, respectively, with complete medium supplemented with 5% human serum or autologous serum in the absence of IL-2. Then, the T cell line is collected, washed to completely remove IL-2 and, after counting, is resuspended in medium without IL-2 at the same concentration of irradiated PBMC. Then the purified antigen used for the first stimulation is given at the same working concentration. This restimulated co-culture is then maintained as described above for primary stimulation.

3.4.2. Restimulation of Antigen-Specific T Cell Clones

To maintain and expand in culture T cell clones, it is necessary to restimulate them every 20–30 days. Since clones represent the progeny of a single antigen-specific cell, i.e., it is not contaminated by other cells that could outgrow, it is not necessary to use the antigen and MHC compatible APC to restimulate them (that are, on the contrary, mandatory for lines). So that, a non-specific stimulus such as the mitogen PHA can be used. T cell clone must be collected, resuspended at 5×10^5 cells/ml (it is not a strict concentration but it should not exceed 1×10^6 cells/ml) and 100 μl or 1 ml placed in 96- or 24-well plates, respectively. Then an equal volume of restimulation mixture (*see* **Section 2**) is added to the culture (*see* **Note** 7).

Check the wells every 2–4 days and as soon as the cells start to grow and the medium turn to yellow, the cells can be split 1:2–1:4, according to the cell expansion, with complete medium supplemented with 5% human serum or 10% FCS and 100 U/ml IL-2. Avoid to keep the cells in acidic yellow medium for more than a day. After 10–15 days the cells slow down the growing rate and therefore they can be split and diluted less often. After restimulation, the T cell clones can be cultured for 3–4 weeks in the presence of IL-2 (*see* **Note 8**).

3.5. Complementary Methods

3.5.1. Isolation of T Lymphocytes From Biopsies or Other Tissues

Clones can be generated from T lymphocytes isolated from tissues *(13, 15)*. However, PBMC must be isolated from the peripheral blood of the same donor. PBMC will be used to obtain an EBV transformed B cell line and frozen to be used as APC in primary and possibly restimulation cultures. Tissue may derive from biopsies of surgery samples and care must be given to the sterile preparation of the tissues. The tissue is to be treated to reduce its mass (in the case of surgery samples) and immersed in culture medium in the presence of antibiotics. Then the tissue must be finely minced under sterile conditions and passed trough a fine plastic sterile net (nylon cell strainer 40 μm) to allow lymphocytes to enter in suspension. In some circumstances, incubation of the minced tissues with enzymes such as collagenase can help to increase the recover of lymphocytes. The cell suspension must be washed twice at low rpm (100) to remove the tissue fragments and the supernatant can be layered onto a Ficoll cushion to isolate lymphocytes. The density gradient isolation may not be required if erythrocytes and other cell contaminants are not relevant and if T cell counting for immediate cloning is not required. In fact, in the majority of cases, the chances to obtain a growing line and clones from tissues increase if cells are not stimulated soon after the isolation from tissues. Since in tissues are more often found cells with effector function and with an activated phenotype due to specific and non-specific stimulation due to the inflammatory environment of the tissue, cells should be maintained in culture for not less that 3–7 days in the medium containing 100 U/ml of IL-2. Then, cells can be stimulated in the presence of the relevant antigen and autologous APC as described for the primary culture. However, the number of cells in this case is usually limited. Thus a proportional number of APC must be calculated and generally a limited number of wells can be programmed. In any case a negative culture in the absence of antigen is required.

3.5.2. Cloning a Non-Antigen-Specific but Rare T Cell Population

A rare T cell population in the blood may be enriched by T cell magnetic- or flow cytometric-sorting. These two methods allow the physical separation of cell subpopulations from a heterogeneous mixture of cell types and rely on the availability of an appropriate monoclonal antibody or a mix of monoclonal antibodies that identify the cell types to be separated. Both these methods can be performed in positive or negative mode utilizing antibodies direct against the required cell population or against all the other cell types, respectively. Their advantages over other antibody-mediated selection techniques are purity of resulting cell preparation and reproducibility of separation. Cell separation using flow cytometry involves scanning and sorting individual cells and generally leads to a cell population of higher purity than that obtained with other techniques but suffers from the disadvantage that the number of cells obtained is limited and time necessary to perform cell-sorting increases proportionally to the purity required. Immuno-magnetic purification of T cells is an easier and less expensive method than the flow cytometric-sorting, since it does not require sophisticated instrumentation and allows handling of larger number of cells, but purity of the recovered populations is reduced in comparison to cytometric-sorting. In both cases, a significant factor limiting the purity of the recovered cell populations is represented by cell clumping, and care must be given to obtain single cell suspension of samples to sort.

3.5.3. Measurement of Proliferative Responses of T Lymphocytes

Measurement of proliferative responses is a fundamental technique for the measurement of T lymphocytes responses to various stimuli. Cell proliferation is determined by estimating the incorporation of ^{3}H-thymidine into DNA during stimuli-induced mitosis. ^{3}H-thymidine can be added after 2 (when using T cell clones and lines) or 6 days of culture (when using PBMC) at 1 μCi/well and left in culture from 6 to 18 h in humidified 37°C, 5% CO_2incubator. A long culture with ^{3}H-thymidine (18 h) yields more reproducible data, but shorter periods (6–8 h) can be used in the case, for example, of screening tests. After the pulsing time, cells have to be harvested using an automated multiwell harvester that aspirates cells, lyses cells, and transfers ^{3}H-thymidine-labeled DNA onto filter paper, while allowing non-incorporated ^{3}H-thymidine in the medium to flow through. Then, after addition of scintillation fluid, the incorporated radioactivity can be measured in a scintillation counter for beta particles as cpm. The resulting data can be analyzed and the SI that represents the ratio of cpm between stimulated and negative control wells calculated. Clones with cpm $>1{,}000$ and SI >3 will be considered antigen-specific.

3.5.4. Block of Cellular Division

The block of feeder cells (APC) cellular division can be performed by gamma-irradiation (3,500 rads for PBMC and 6,000 rads for BCLL) with a cobalt source immediately before use. After irradiation, wash three times the cells in order to remove possible toxic substances released during irradiation.

In some laboratories a cobalt source is not available, thus, the block of APC cellular division can be obtained treating cells with mitomycin C (a DNA cross-linking reagent). Wash the cells two times in RPMI without serum and resuspend them with 2 ml of RPMI. Then, add 100 μl of mitomycin C at 1 mg/ml (final concentration: 50 μg/ml) and incubate 1 h at 37°C on the dark. After incubation time, wash extensively the cells (at least four times) to remove any trace of mitomycin C, which could be toxic for the responder T cells.

3.5.5. EBV Transformed B Cell Line BCLL

Immortalization of B lymphocytes by EBV is a procedure for inducing long-term growth of human B lymphocytes *(26)*. EBV may cause disease in non-immune individuals, therefore biosafety practices must be followed. The basic protocol to obtain a BCLL can be divided into two steps:

1. Preparation of EBV-containing culture supernatants: resuspend 1×10^6 cells/ml of exponentially growing B95.8 cell line in complete medium supplemented with 10% FCS and incubate 3 days in humidified 37°C, 5% CO_2 incubator. Then collect the culture supernatant, filter, and store at –80°C aliquots until use.
2. EBV infection and growth of infected B cells: 1×10^7 PBMC are incubated with EBV-containing supernatant in the presence of 500 ng/ml of cyclosporine A in complete medium containing 10% FCS. Cells are plated on 96-well plates and kept incubated at 37°C, 5% CO_2 for 3–4 weeks. Every week, 2 ml of culture medium is removed and 2 ml of fresh medium added until the growth of BCLL is established.

3.5.6. Freezing T Cell Lines and Clones

As with any technique that requires long-term cell culture with frequent manipulation of the various cultures, contamination with subsequent loss of valuable cell lines is the most common problem encountered in T cell cloning. Thus, it is important to freeze several aliquots of each cell line and clone (as soon as sufficient numbers of cells became available) for future use and when a T cell clone or line is not necessary for experiments. To recover cells in good condition upon thawing, T cell lines and clones are frozen after 5–7 days from the last stimulation. The cells are counted, pelleted, and resuspended up to 5×10^6cells/ml in freezing medium (*see* **Section 2**). Aliquots of 0.5–1 ml are prepared in cryotubes and transferred immediately into a –70°C freezer. This intermediate –70°C incubation is important to inhibit any metabolic activity of the cells in the presence of DMSO. Moreover, to improve the vitality of cells at thawing by avoiding the formation of crystals within the cells, cryovials should be placed into "freezing can" containing ethanol or into a polystyrene box. These procedures cause the temperature of the cells within the cryovials to decrease linearly. Frozen cell lines and clones are stable at –70°C for approximately a week (recovered viability slowly decreases over a period of 4 weeks). Conservation of cells for longer period is possible if cryovials are transferred into liquid nitrogen containers. The concentration of cells in each aliquot may vary: in general, a range of 1×10^6–1×10^7 T cells per aliquot is adequate to obtain a good recovery after thawing.

3.5.7. Generation of Protein Antigen-Specific CD8^{+ve} T Cell Clones

The procedures to obtain CD8+ T cell clones are in theory identical to those used for the generation of antigen-specific CD4+ clones. However, two main differences must be taken into consideration as well as the different procedures for evaluating the specificity. The differences are related to the different antigens that stimulate CD8+ T cells and to the antigen processing pathway that allows a given peptide to bind MHC class I. The majority of soluble antigen, when picked-up by APC is internalized by phagocytosis–endocytosis and targeted to phagosomes-late endosomes, where they intersect the MHC class II pathway so that digested peptides are associated to these HLA molecules. Consequently, MHC class II specific CD4+ T cells are likely to be activated. The generation of a CD8+ T cell line can be obtained using peptides, which directly bind MHC class I. Peptides are synthesized according to the antigen sequence and the MHC class I binding motifs *(27–29)*. Alternatively, a viral vector encoding for the wanted protein antigen can be used. Moreover, the capacity of dendritic cells (DC) to cross-present antigens within apoptotic cells or exosomes can be exploited *(30, 31)*. Lastly, methods to induce the soluble antigen migration into the cytoplasmatic compartment have been adopted, such as the use of endosomal pH reducing antigen (chloroquine) *(32)*. When the peptide antigen is known, the tetramer technology *(33)* can be used to identify and sort the specific T cell population to expand and finally clone by limiting dilution. The same strategy used to identify and expand the specific T cells can be used to screen for the specific cells after cloning, using as read-out a cytotoxicity assay or cytokine secretion assay.

3.5.8. Generation of CD1-Restricted T Cell Clones

The same procedures described for cloning of CD4 peptide-specific T cells, with some modifications, can be used to obtain CD1-restricted T cell clones. In particular, since CD1+ APC are required, the main differences are limited to the set-up of the primary culture and screening tests. To optimize the possibility to expand lipid-specific CD1-restricted T cells instead of peptide-reactive MHC class I- or II-restricted T lymphocytes, the primary culture must be set-up using DC as APC. In details, from a given donor, PBMC must be isolated twice at a 6 days interval. Obviously, from a single blood withdrawal PBMC can be isolated and frozen at least two aliquots. From the first PBMC isolation, monocytes are magnetically sorted and induced to differentiate into DC *(34)*. After 5 days, DC are pulsed overnight with the purified lipid antigen of interest in serum free complete medium, to allow internalization and surface expression of antigen–CD1 complexes. The pulsing in serum free medium is a procedure that, avoiding the interference by serum lipids, increases the CD1-loading with the given lipid antigen. At the 6th day, a new isolation of PBMC (or thawing of a PBMC aliquot) from the same donor is required as CD1-restricted T lymphocyte source. Then, the o/n pulsed DC can be co-cultured in complete medium together with autologous PBMC (ratio DC:PBMC=1:10) and 5% serum, in the presence of blocking anti-MHC class I and class II monoclonal antibodies (at 5 μg/ml). The next steps of primary culture expansion and cloning are identical to the procedure related to peptide-specific T cell clone generation described above .

The screening of the antigen-specific CD1-restricted T cell clones among all the growing clones presents some differences with the MHC class II-restricted T cell clone screening. First of all, the source of APC, in this case, is represented by DC, thus thawing of a further aliquot of PBMC (or a new isolation of PBMC from the same donor) is required 4–7 days before the screening test to induce monocyte differentiation into DC. Then, instead of the o/n pulsing of APC, the lipid antigen is given to DC at the optimal dose and left in the culture of APC and responder T clones. In particular, two aliquots of DC are prepared: one without and the other with the lipid antigen. After a 2 h incubation in serum free medium at 37°C, to allow the loading of CD1 molecules with the lipid antigen, T cell clones are added with a DC:T ratio 1:10. Lastly, autologous DC are of help in the screening, but DC from any donor can be used in all other lipid antigen-specific clones test, since CD1 are non-polymorphic molecules. A typical screening includes a first test with autologous DC, to screen for all the antigen-specific T cell clones regardless of restriction element. Then, a second antigen presentation assay can be performed using diverse sources of allogeneic DC to identify the CD1-restricted lipid antigen-specific clones. As a last confirmation test, an antigen presentation assay in the presence of anti MHC class I and II and anti CD1 monoclonal antibodies (at 5 μg/ml) will identify the restriction element *(35, 36)*.

Supplementary materials:

GM-CSF

IL-4

Anti MHC class I monoclonal antibody (i.e., W6-32, ATCC)

Anti MHC class II monoclonal antibody (i.e., L243, ATCC)

3.5.9. Preparation of Autologous Serum

To obtain autologous serum, venous blood is collected in the absence of anticoagulant and kept into a sterile tube for 1 h at 37°C to allow clotting (10 ml of blood from the same donor of the PBMC will give roughly 4 ml of serum). Then, place the tube at 4°C for 1 h. Using a sterile stick, gently detach the clot from the wall of the tube, paying attention not to break the clot itself. Then, centrifuge the tube 2,500 rpm at 4°C and collect the supernatant, that can be frozen until use. Avoid repeated freezing/thawing cycles of sera.

4. Notes

1. A constrain in the generation of human antigen-specific T cell clones is the availability of a purified antigen. Several antigens can be purchased from the market, such as TT, or isolated from a given pathogen or cell or chemically synthesized. In general, the more purified is the antigen, the higher will be the chance to obtain a population of specific T cell clones. For example, a recombinant protein produced in *Escherichia coli* would represent an excellent source of antigen. However, care must be given to highly purify the recombinant protein from products of the bacterial cells in which the protein has been synthesized. Since every human being has daily contacts with *E. coli*, the frequency of T cells specific for *E. coli* proteins is high in the peripheral blood of almost all the individuals. Thus, the culture of T lymphocytes with a recombinant protein produced in *E. coli* without a high purification from bacterial products has high probability to give rise to a T cell line specific for *E. coli* antigen possibly "contaminated" by T cells specific for the recombinant protein. In any case, if a recombinant protein is to be used as source of antigen, a protein extract from the vector will be required to exclude unwanted specificities in the antigen-responsiveness tests that have been described above.
2. When the concentration of the antigen and the expected response is known, for instance, in the case of the generation of clones specific for the same antigen from a second donor after having isolated clones from a first subject, we could design a more simple primary culture experiment. For example, if we have defined that the best concentration for TT to be used is 1 μg/ml and that recently vaccinated subjects have a high response so that in the first experiment a high number of 96 wells with this dose of antigen end-up as positive, we could simply prepare a primary culture by seeding cells at 5×10^6/ml in 24-well plates filled with 2 ml of complete medium supplemented with 5% autologous serum. We could prepare two

wells without antigen, as negative control, and from two to five wells with antigen at the previously defined dose. IL-2 will be added at the same concentrations indicated above at day 5 and 10. If needed, wells could be split in neighbor wells using the same medium, until cloning.

3. PBMC can be frozen after irradiation and used afterwards for the preparation of the restimulation mixture. Cloning mixture, on the other hand, should be prepared using irradiated PBMC after isolation from blood.

4. Since the volume in each well is limited to 20 μl in Terasaki plates, it may happen that in the 12 days of culture most of the medium evaporates. This will cause the death of the cells. To avoid evaporation of the well content, it is preferable to use an incubator that is not utilized for the culture of other cells, so that you can avoid opening it for the all culture period. It is also suggested to add an extra 2 l of sterile water in a flat and opened plastic or metallic container to increase the humidity in the incubator. In addition, the plates must not be microscopically inspected until the 12th day. Inspection under the microscope will increase the evaporation rate and the number of dry wells will increase, accordingly.

5. Since a cloned population of T cells has the same antigen specificity, all the culture procedure that follows the cloning procedure in Terasaki or other plates can be performed using FCS instead the more expensive and more difficult to handle human serum or the non always available autologous serum. The use of human serum can be suggested for clones that do grow slowly.

6. Screening assays of many clones may be complex, since IL-2 in the culture must be washed out from every clone culture. It is possible to wash many clones at contemporaneously by the use of a refrigerated centrifuge with a rotor allowing the 96 wells microplate canister. In the plate for antigen presentation test, each clone should be moved in two wells: one to be challenged with APC without antigen, the other with APC pulsed with antigen. Since clones to be screened are growing in 96-well plates in 200 μl of medium, a simple procedure will permit their transfer to plates for the antigen specificity test. Using a 8-channel micropipette, each clone can be transferred in the plate for the antigen presentation test with a procedure similar to that depicted in **Fig. 6.4**. In details, from each vertical row of the growing plate, move 60 μl of clone to the first and second rows of the test plate. In this case, in the culture plate approximately 80 μl of each clone will remain for further culture and restimulation. When this procedure will be repeated for all the clones with the same multipipette 140 μl of medium is added

to each clone in the test plate and 120 μl of complete medium +IL-2 is added to the growing culture plates. The plate for growing is moved to the incubator for culture, while the test plates are spun in the centrifuge and clones washed twice with medium without IL-2.

The multipipette can be used to change the medium for washings, paying attention to not move or aspirate the cells in the bottom of the wells. After washings using the multipipette 100 μl of medium is added and then 100 μl of the APC with or without antigen in the appropriate rows.

7. As indicated in **Section 2**, irradiated PBMC from any healthy donor can be used. Since the cells of a clone cannot change/rearrange/mutate their TCR, the antigen specificity is maintained also if PHA is used to induce their proliferation and consequently there is no MHC restriction.
8. Since upon stimulation T lymphocytes downregulate their TCR and proliferate/secrete cytokine irrespective of further stimuli, it is not possible to use clones for antigen specificity tests before 7–8 days from the restimulation. Moreover, since clones express IL-2 receptor, in any specificity test IL-2 should not be added to avoid high-background proliferation/cytokine secretion and non-reliable results.

References

1. Lanzavecchia A, Sallusto F. Dynamics of T lymphocyte responses: intermediates, effectors, and memory cells. Science 2000;290:92–7.
2. Lanzavecchia A. Understanding the mechanisms of sustained signaling and T cell activation. J Exp Med 1997;185:1717–9.
3. Colombetti S, Basso V, Mueller DL, Mondino A. Prolonged TCR/CD28 engagement drives IL-2-independent T cell clonal expansion through signaling mediated by the mammalian target of rapamycin. J Immunol 2006;176:2730–8.
4. Saha K, Volsky DJ, Matczak E. Resistance against syncytium-inducing human immunodeficiency virus type 1 (HIV-1) in selected CD4(+) T cells from an HIV-1-infected nonprogressor: evidence of a novel pathway of resistance mediated by a soluble factor(s) that acts after virus entry. J Virol 1999;73:7891–8.
5. Maynard J, Petersson K, Wilson DH, et al. Structure of an autoimmune T cell receptor complexed with class II peptide-MHC: insights into MHC bias and antigen specificity. Immunity 2005;22:81–92.
6. Artini M, Nisini R, Missale G, et al. Infection of circulating and liver infiltrating T cells by hepatitis C virus of different subtypes. Viral Immunol 1995;8:63–73.
7. Kelso A, Groves P, Troutt AB, Francis K. Evidence for the stochastic acquisition of cytokine profile by CD4+ T cells activated in a T helper type 2-like response in vivo. Eur J Immunol 1995;25:1168–75.
8. Demotz S, Lanzavecchia A, Eisel U, Niemann H, Widmann C, Corradin G. Delineation of several DR-restricted tetanus toxin T cell epitopes. J Immunol 1989;142:394–402.
9. Lanzavecchia A, Santini P, Maggi E, et al. In vitro selective expansion of allergen specific T cells from atopic patients. Clin Exp Immunol 1983;52:21–8.
10. Romagnani S, Maggi E, Parronchi P, Macchia D, Piccinni MP, Ricci M. Increased numbers of Th2-like CD4+ T cells in target organs and in the allergen-specific repertoire of allergic patients. Possible role of IL-4 produced by non-T cells. Int Arch Allergy Appl Immunol 1991;94:133–6.

11. Nikolova M, Echchakir H, Wechsler J, Boumsell L, Bensussan A, Bagot M. Isolation of a CD8alphaalpha+ CD4- tumour T-cell clone with cytotoxic activity from a CD4+ CD8- cutaneous T-cell lymphoma. Br J Dermatol 2003;148:24–9.
12. Sobel DO, Creswell K. Characterization of anti-islet cytotoxic human T-cell clones from patients with type 1 diabetes mellitus. Autoimmunity 2006;39:323–32.
13. Franco A, Barnaba V, Ruberti G, Benvenuto R, Balsano C, Musca A. Liver-derived T cell clones in autoimmune chronic active hepatitis: accessory cell function of hepatocytes expressing class II major histocompatibility complex molecules. Clin Immunol Immunopathol 1990;54:382–94.
14. Wang RF, Appella E, Kawakami Y, Kang X, Rosenberg SA. Identification of TRP-2 as a human tumor antigen recognized by cytotoxic T lymphocytes. J Exp Med 1996;184:2207–16.
15. Gervois N, Guilloux Y, Diez E, Jotereau F. Suboptimal activation of melanoma infiltrating lymphocytes (TIL) due to low avidity of TCR/MHC-tumor peptide interactions. J Exp Med 1996;183:2403–7.
16. Nisini R, Paroli M, Accapezzato D, et al. Human CD4+ T-cell response to hepatitis delta virus: identification of multiple epitopes and characterization of T-helper cytokine profiles. J Virol 1997;71:2241–51.
17. Unutmaz D, Pileri P, Abrignani S. Antigen-independent activation of naive and memory resting T cells by a cytokine combination. J Exp Med 1994;180:1159–64.
18. Morley AA, Trainor KJ, Seshadri RS. Cloning of human lymphocytes using limiting dilution. Exp Hematol 1983;11:418–24.
19. Sharrock CE, Kaminski E, Man S. Limiting dilution analysis of human T cells: a useful clinical tool. Immunol Today 1990;11:281–6.
20. Nisini R, Fattorossi A, Ferlini C, D'Amelio R. One cause for the apparent inability of human T cell clones to function as professional superantigen-presenting cells is autoactivation. Eur J Immunol 1996;26:797–803.
21. Nisini R, Matricardi PM, Fattorossi A, Biselli R, D'Amelio R. Presentation of superantigen by human T cell clones: a model of T-T cell interaction. Eur J Immunol 1992;22:2033–9.
22. Nisini R, Romagnoli G, Gomez MJ, et al. Antigenic properties and processing requirements of 65-kilodalton mannoprotein, a major antigen target of anti-Candida human T-cell response, as disclosed by specific human T-cell clones. Infect Immun 2001;69:3728–36.
23. Bonnefoix T, Bonnefoix P, Mi JQ, Lawrence JJ, Sotto JJ, Leroux D. Detection of suppressor T lymphocytes and estimation of their frequency in limiting dilution assays by generalized linear regression modeling. J Immunol 2003;170:2884–94.
24. Waldmann H, Pope H, Lefkovits I. Limiting dilution analysis of helper T-cell function. II. An approach to the study of the function of single helper T cells. Immunology 1976;31:343–52.
25. Waldmann H, Lefkovits I, Quintans J. Limiting dilution analysis of helper T-cell function. Immunology 1975;28:1135–48.
26. Pelloquin F, Lamelin JP, Lenoir GM. Human B lymphocytes immortalization by Epstein-Barr virus in the presence of cyclosporin A. In Vitro Cell Dev Biol 1986;22:689–94.
27. Parker KC, Bednarek MA, Hull LK, et al. Sequence motifs important for peptide binding to the human MHC class I molecule, HLA-A2. J Immunol 1992;149:3580–7.
28. Sette A, Buus S, Appella E, et al. Prediction of major histocompatibility complex binding regions of protein antigens by sequence pattern analysis. Proc Natl Acad Sci USA 1989;86:3296–300.
29. Stevanovic S, Jung G. Multiple sequence analysis: pool sequencing of synthetic and natural peptide libraries. Anal Biochem 1993;212:212–20.
30. Utsugi-Kobukai S, Fujimaki H, Hotta C, Nakazawa M, Minami M. MHC class I-mediated exogenous antigen presentation by exosomes secreted from immature and mature bone marrow derived dendritic cells. Immunol Lett 2003;89:125–31.
31. Lapenta C, Santini SM, Spada M, et al. IFN-alpha-conditioned dendritic cells are highly efficient in inducing cross-priming CD8(+) T cells against exogenous viral antigens. Eur J Immunol 2006;36:2046–60.
32. Accapezzato D, Visco V, Francavilla V, et al. Chloroquine enhances human CD8+ T cell responses against soluble antigens in vivo. J Exp Med 2005;202:817–28.
33. Hoffmann TK, Donnenberg VS, Friebe-Hoffmann U, et al. Competition of peptide-MHC class I tetrameric complexes with anti-CD3 provides evidence for specificity of peptide binding to the TCR complex. Cytometry 2000;41:321–8.

34. Sallusto F, Lanzavecchia A. Efficient presentation of soluble antigen by cultured human dendritic cells is maintained by granulocyte/macrophage colony-stimulating factor plus interleukin 4 and downregulated by tumor necrosis factor alpha. J Exp Med 1994;179: 1109–18.
35. Gilleron M, Stenger S, Mazorra Z, et al. Diacylated sulfoglycolipids are novel mycobacterial antigens stimulating CD1-restricted T cells during infection with Mycobacterium tuberculosis. J Exp Med 2004;199:649–59.
36. Russano AM, Agea E, Corazzi L, et al. Recognition of pollen-derived phosphatidyl-ethanolamine by human CD1d-restricted gamma delta T cells. J Allergy Clin Immunol 2006; 117:1178–84.

Chapter 7

Limiting Dilution Analysis of Antigen-Specific T Cells

Jorge Carneiro, Lurdes Duarte, and Elisabetta Padovan

Abstract

Limiting dilution analysis (LDA) has been extensively employed as a quantitative method to estimate the precursor frequency of various T lymphocyte subsets according to their functional properties in vitro. We describe here an example of LDA experiment assessing antigen-specific T cell proliferation of microcultures in the presence or absence of adjuvant and illustrate how to estimate the frequencies of precursor T cells using an online tool that we made publicly available.

Key words: Antigen, antigen presenting cells, $CD4^+$ T cells, cell proliferation, human, limiting dilution, microculture, single-hit/multi-hit Poisson model.

1. Introduction

In recent years several methods have been employed to estimate the precursor frequency of various T lymphocyte subsets. A summary of most used assays is presented in **Table 7.1**. Induction of cytokines in response to an antigen-specific stimulation is a general method to assess $CD4^+$ and $CD8^+$ T cell responses in vitro, using ELISPOT or FACS analysis. FACS analysis upon labeling of lymphocytes with tetramers or CFSE is largely employed to enumerate Ag-specific T cells in vitro and ex vivo. Although these assays can detect expansion of precursor T cells present at very low frequencies ($1/10^6$), the cellular responses measured are representative of neither $CD8^+$ T cell functions, nor $CD4^+$ T lymphocyte activity, unless B cell responses are measured in the same assay. Furthermore, the frequency of Ag-specific T cells measured with ELISPOT and FACS staining methods are overestimated *(1, 2)*. LDA represents instead a quantitative method to determine T cell

Gennaro De Libero (ed.), *T Cell Protocols: Second Edition, vol. 514*

DOI 10.1007/978-1-60327-527-9_7 Springerprotocols.com

Table 7.1
Characteristics of cytokine detection, tetramer staining, CFSE staining and LDA methods used for the determination of Ag-specific T cell precursor frequency

	Cytokine	Tetramer	CFSE staining	LDA
Lymphocyte subset	$CD4^+$ and $CD8^+$ T cells	$CD8^+$ T cells	$CD4^+$ and $CD8^+$ T cells	$CD4^+$ and $CD8^+$ T cells
Read-out	ELISPOT and FACS staining	FACS staining	FACS staining	^{3}HTdR incorporation and ^{51}Cr-release
Cellular response	Production of cytokines	Expression of Ag-specific TCR	Cell proliferation	Cell proliferation and lytic activity
Requirement	Need of secondary non-specific T cell stimulation	Suitable for ex-vivo and in vitro studies	Suitable for ex-vivo and in vitro studies	Need of T cell expansion in vitro
Detection range	$1/10^3$–$1/10^6$	$1/10^2$–$1/10^5$	$1/10^2$–$1/10^5$	$1/10^4$–$1/10^5$

precursor frequencies based on lymphocyte functional properties *(3)*. Incorporation of ^{3}H-thymidine is employed to measure proliferation of $CD4^+$ T helper cells microcultures, while release of ^{51}Cr from $Na_2{}^{51}CrO_4$-labeled target cells is used to quantify the lytic activity of cytotoxic $CD8^+$ T cells seeded in limiting dilution conditions. LDA requires first the outgrowth of T cells in culture, and then a functional read-out, thus allowing the detection of functional lymphocyte precursors in the range of $1/10^4$–$1/10^5$.

We will make use of the effect of synthetic compounds known to act as immune enhancers on antigen-specific T lymphocytes *(4, 5)* to illustrate the setting of a classical experiment of LDA analysis for $CD4^+$ T helper cells and show how to determine the cell precursor cell frequencies, using the single-hit and multi-hit Poisson model *(3)*.

2. Materials

1. Peripheral blood mononuclear cells (PBMC) isolated from human healthy donors.
2. Complete cell culture medium: RPMI 1640 containing 1% sodium pyruvate, non-essential amino acids, glutamax and

kanamycin, 0.1% β-mercaptoethanol (Invitrogen, Basel, CH) and 5% human AB serum (*see* **Note 1**).

3. Medium for cell washing: RPMI 1640 containing 10 mM Hepes and 0.5% human AB serum.
4. The synthetic lipopeptide adjuvant P_3CSK_4 and the CSK_4 control carrier peptide (EMC microcollections, Tübingen, D).
5. The antigenic peptide $HBsAg_{19-33}$ (FFLLTRILTIPQSLD), which binds promiscuous HLA-DR molecules *(6)*.
6. Tritiated (^3H)-thymidine at 50 μCi/ml prepared by adding 1 ml of ^3H-thymidine at 600 μCi/ml (TRK758; Amersham Biosciences, Little Chalfont, UK) to 19 ml of cell culture medium.
7. A cell harvester and a β-counter. We harvested the cells on GF/A glass fiber filters and measured cell-associated ^3H-thymidine with an automatic β-counter provided by Inotech, Asbach, D.
8. Cell culture plates with 24 flat-bottom and 96 U-bottom wells.
9. A cell centrifuge.
10. A cell incubator at 37°C and 5% CO_2.
11. A γ-irradiator.
12. Variable volume pipettes and a 12-channel pipette.

3. Methods

Below we describe (1) how to prepare and expand a primary culture of (Ag)-specific T lymphocytes in vitro, (2) how to restimulate the generated polyclonal T cell line with the specific Ag in limiting dilution condition, (3) how to measure proliferation of T cell microcultures, and (4) how to choose a single-hit or multi-hit Poisson model and to determine the frequency of Ag-specific T cells.

3.1. T Cell Priming In Vitro

1. Antigen-specific T helper cell lines are best generated from freshly isolated PBMC (*see* **Note 2**).
2. Total PBMC are diluted in cell culture medium at 1.5×10^6 cells/ml and 1 ml of the cell suspension is seeded in two wells of a 24-well plate.
3. The first well is then stimulated with 1 ml of $HBsAg_{19-33}$ peptide diluted in cell culture medium at 6 μg/ml in the presence of CSK_4 at 200 ng/ml.

4. A mixture of $HBsAg_{19-33}$ peptide at 6 μg/ml and P_3CSK_4 at 200 ng/ml is added to the second well containing PBMC. Each well will thus contain 1.5×10^6 PBMC, primed with 3 μg/ml of Ag and 100 ng/ml of adjuvant/carrier in a final volume of 2 ml.
5. In order to harvest sufficient cells for the limiting dilution cultures described in **Section 3.2**, we recommend seeding multiple wells for each priming condition.
6. The remaining PBMC can be kept frozen (*see* **Note 3**) and used as antigen presenting cells (APC) in the limiting dilution cultures.
7. Primed lymphocytes are incubated at 37°C in 5% CO_2 for 5 days.

3.2. Stimulation of Primed Lymphocyte in Limiting Dilution Condition

1. After 5 days of priming, cells are restimulated in limiting dilution conditions. To this aim, autologous Ag-pulsed/non-pulsed and irradiated APC are needed.
2. To prepare Ag-pulsed APC, autologous PBMC are diluted at $5–10 \times 10^6$ /ml in cell culture medium containing 10 mM Hepes, $HBsAg_{19-33}$ 3 μg/ml and either adjuvant or carrier at 100 ng/ml.
3. To prepare non-pulsed APC, a smaller aliquot of PBMC is diluted in a separate tube with either adjuvant or carrier at 100 ng/ml, without antigen.
4. All APC-containing tubes are placed at 37°C for 1–2 h; tubes are gently shaken every 10 min.
5. Lymphocyte cultures primed in the presence of antigen and adjuvant/carrier are then harvested.
6. Cultures set in the same priming conditions may be pooled together.
7. Cells are washed once and diluted at 2×10^5 cells/ml in complete cell culture medium.
8. Cell mixtures of 200 μl/well are seeded in a 96 U-bottom cell plates (plate A). In our example, we have seeded a total of 36 wells per culture, however, seeding a higher number of wells, will reduce the experimental error.
9. To prepare serial dilution of primed lymphocytes, 100 μl/well of cell culture medium are first added in 36 wells of three empty 96 U-bottom plates (plates B, C, D).
10. Using a 12-channel pipette, 100 μl/well of the cell suspension is harvested from the first row of plate A and transferred into the first row of plate B, then in the first row of plate C, and finally into the first row of plate D.
11. The excess of 100 μl/well of cell suspension from plate D is discarded.

12. The procedure is repeated for each row of plate A through plate D. When the transfer is completed, each well (or microculture) contains 100 μl/well of cell culture medium, with serial dilution of cells: plate A, 2×10^4 cells/well; plate B, 10^4 cells/well; plate C, 5×10^3 cells/well; plate D, 2.5×10^3 cells/well.
13. Microcultures can be placed in the incubator while preparing the APC.
14. Ag-pulsed and non-pulsed PBMC are irradiated at 3,500 rad and washed two times, changing tube once.
15. After the last wash, Ag-pulsed and irradiated PBMC are diluted at 5×10^5/ml in cell culture medium and seeded in 100 μl/well into 33 of the 36 microcultures contained in plates A–D.
16. Non-pulsed and irradiated PBMC are diluted in the same conditions and seeded in 100 μl/well into the remaining three wells of each plate.
17. When completed, each plate contains a total of 36 microculture with 5×10^4 PBMC/well and primed lymphocyte at a given serial dilution in a final volume of 200 μl/well.
18. The three microcultures with non-pulsed irradiated PBMC represent the negative control of stimulation.
19. All culture plates are incubated for 48 h at 37°C in 5% CO_2.

3.3. Determination of T Cell Proliferation

1. After 2 days, 10 μl/well of ^{3}H-thymidine at 50 μCi/ml are added to each microculture and incubated at 37°C in 5% CO_2 for 16–18 h.
2. Microcultures are finally harvested on filter papers and cell-associated radioactivity measured with an automatic β-counter.
3. Raw experimental data are given as cpm (counts per minute) for each single microcultures as represented in **Tables 7.2** and **7.3**.

3.4. LDA Analysis

LDA data from T cell cultures are first analyzed using the single-hit Poisson model assuming that only one cell is sufficient for generating a positive response (*see* **Note 4**). Since the application of this model requires the determination of the number of microcultures that scored negative in proliferation, we first determined the threshold of proliferation indicative of positive responses on the basis of the results obtained in all microcultures stimulated with non-pulsed and irradiated PBMC, irrespective of the limiting dilution (*see* **Note 5**). As illustrated in **Table 7.2**, we set this threshold at a value of at least 3 SD above the mean of the proliferation in the absence of antigen. Thus, cultures in the

Table 7.2
Proliferative responses of limiting dilution cultures in the presence of non-pulsed, irradiated PBMC

	^{3}H-Thymidine incorporation (cpm)							
Stimuli	CSK4				P3CSK4			
Cells/well	20,000	10,000	5,000	2,500	20,000	10,000	5,000	2,500
	804	198	44	22	2,250	417	40	225
	831	168	67	117	1,640	446	160	82
	290	343	75	35	1,477	555	160	62
Mean	250	–	–	–	626	–	–	–
SD	284	–	–	–	740	–	–	–
Mean+(SD*3)	1,101	–	–	–	2,846	–	–	–

Table 7.3
Proliferative responses of limiting dilution cultures in the presence of Ag-pulsed, irradiated PBMC

	^{3}H-Thymidine incorporation (cpm)							
Stimuli	HBsAg + CSK4				HBsAg + P3CSK4			
Cells/well	20,000	10,000	5,000	2,500	20,000	10,000	5,000	2,500
	1,380	398	280	146	7,297	1,412	831	572
	516	658	292	442	4,484	5,004	918	280
	1,763	538	322	297	5,295	2,545	1,941	360
	1,771	303	325	115	6,844	4,353	523	386
	904	818	477	241	4,542	2,480	538	297
	721	565	107	215	5,250	3,149	635	335
	866	1,400	125	410	3,589	3,051	1,261	297
	1,816	843	580	122	5,419	1,659	668	1,092
	791	366	274	455	3,900	3,389	918	406
	1,494	883	223	300	7,454	2,314	370	528
	579	668	105	112	5,918	2,859	1,665	173

Table 7.3 (continued)

	^{3}H-Thymidine incorporation (cpm)							
Stimuli	**HBsAg + CSK4**				**HBsAg + P3CSK4**			
Cells/well	**20,000**	**10,000**	**5,000**	**2,500**	**20,000**	**10,000**	**5,000**	**2,500**
	356	333	350	392	4,623	2,420	1,219	293
	1,636	936	2,574	57	1,996	1,374	788	289
	1,722	918	113	292	6,129	1,323	1,412	383
	1,172	405	188	20	4,906	427	1,267	773
	2,391	469	230	92	5,113	862	1,058	755
	958	1,161	255	147	5,272	3,460	293	550
	1,354	871	233	117	6,779	1,960	288	403
	1,801	1,181	443	485	5,822	2,580	501	2,146
	2,254	668	205	225	10,680	3,088	240	348
	979	811	815	265	5,258	3,962	503	408
	980	726	253	580	2,249	2,262	706	242
	1,035	415	358	177	2,140	983	1,093	265
	910	245	510	415	2,831	730	656	527
	2,559	911	235	382	1,632	1,913	358	252
	1,118	1,714	648	60	3,582	449	703	295
	947	1,306	208	52	4,302	1,542	245	430
	1,343	490	888	85	2,672	2,821	1,206	240
	1,629	313	333	47	3,650	484	258	252
	1,158	776	463	404	4,007	1,409	1,216	267
	1,857	448	965	190	2,710	2,645	868	969
	1,449	1,449	255	42	2,608	1,048	1,139	395
	877	671	610	240	4,241	950	581	329
Nr. of wells below 1,101 cpm	14	27	32	33	–	–	–	–
Nr. of wells below 2,846 cpm	–	–	–	–	8	24	33	33

presence of CSK_4 are considered positive if exceeding 1,101 cpm, while cultures grown in the presence of P_3CSK_4, are positive if scoring above 2,846 cpm.

Once the thresholds are defined, the number of negative microcultures is quickly established, as illustrated in **Table 7.3**.

Knowing the cell input per microculture (20,000, 10,000, 5,000, 2,500), the number of microculture seeded (33) and the number of negative wells (CSK_4: 14, 27, 32, 33; P_3CSK_4: 8, 24, 33, 33) for each priming condition, it is possible to estimate the goodness of fit of the single-hit model and calculate the frequency of Ag-specific T cells using the online tool that we made publicly available at: http://eao.igc.gulbenkian.pt/ti/Soft/lda/index.html

The graphical representation of this LDA analysis illustrated in **Fig. 7.1**, reveals that the frequency of HBsAg-specific lymphocytes in the presence of CSK_4 and P_3CSK_4 would be 3.5×10^{-4} and 1.9×10^{-4}, respectively, according to the single-hit model.

However, the plots in **Fig. 7.1** also indicate that the plotted data do not strictly follow a single-hit Poisson curve. If so, the graphical representation would match the straight-line. Following this diagnosis, the data are re-analyzed using a multi-hit model (*see* **Note 6**). The best fitting of the multi-hit model, shown in **Fig. 7.2**, indicates that the frequency of HBsAg-specific lymphocytes in the presence of CSK_4 and P_3CSK_4 is 6.7×10^{-4} and 3.9×10^{-4}, respectively, and that measuring a proliferative response requires at least three and four cells, respectively. Comparison of the single-hit and multi-hit estimates indicates that by taking the former single-hit estimates one would have incurred in errors, although the qualitative order of the estimates is the same.

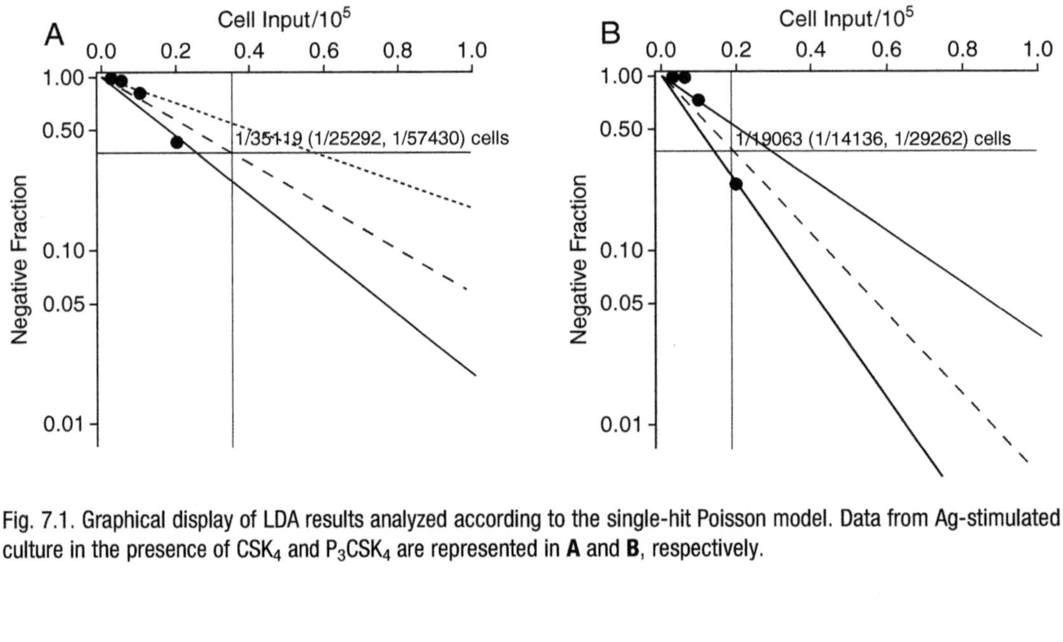

Fig. 7.1. Graphical display of LDA results analyzed according to the single-hit Poisson model. Data from Ag-stimulated culture in the presence of CSK_4 and P_3CSK_4 are represented in **A** and **B**, respectively.

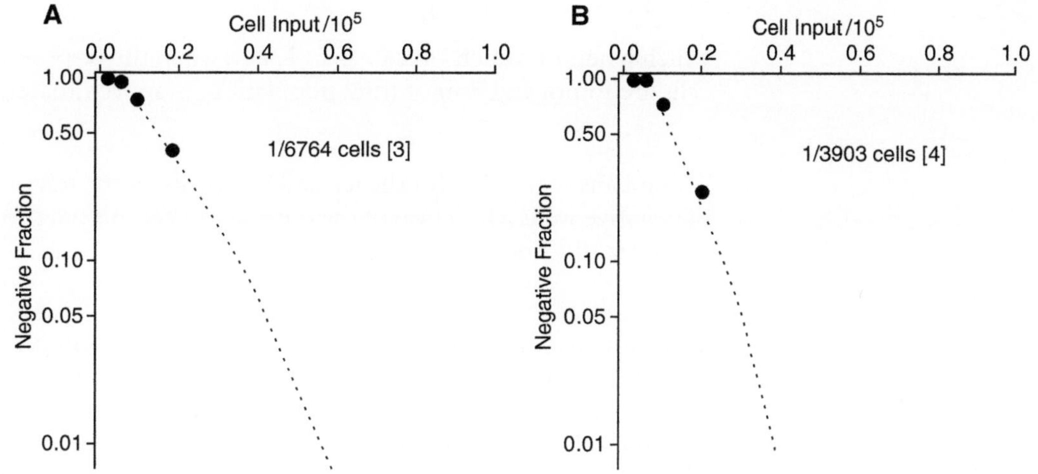

Fig. 7.2. Graphical display of LDA results analyzed according to the multi-hit Poisson model. Data from Ag-stimulated culture in the presence of CSK_4 and P_3CSK_4 are represented in **A** and **B**, respectively.

4. Notes

1. The quality of human serum strongly influences lymphocyte viability and growth in vitro. In our experience, best culture conditions are obtained using human sera provided by the blood bank of local Hospitals.
2. If frozen PBMC were to be employed, de-freeze the vial at 37°C using a water bath. Transfer the cells in a 15 ml cell culture tube and fill it drop-by-drop with washing medium. Centrifuge the tube at 1,200 rpm for 10 min, discard the supernatant, dilute the cell pellet in washing medium and repeat the centrifugation. Proceed as described in **Section 3.1**.
3. To freeze remaining PBMC, wash cells once and dilute the pellet in freezing medium (FBS 40%, DMSO 10% in RPMI 1640) at a density of 5–50 × 10^6/ml. Prepare freezing vials containing 1 ml of cell mixture, freeze the vials at –80°C in a stiropore box and transfer in liquid nitrogen after 24–48 h.
4. The quantitative description of limiting dilution data is typically analyzed using single-hit Poisson model. This model assumes that a single responder cell in the population sample seeded into culture well is sufficient to produce a functional measurement. In our case, this means that the progeny of a responder single cell can incorporate enough thymidine to be detected as above background cpm measurement. Under these conditions, we consider the average probability that none of the c cells seeded into the well is a responding cell, when the frequency of responder cells is r. This probability is:

$$F_o = (1 - r)^c$$

which when r is much smaller than 1, i.e., when the responder cells are minor fraction of total population, is approximately:

$$F_o = e^{-rc}$$

This means that if we plot the natural logarithm of the fraction of negative wells a function of the number of seeding cells c we get a straight line:

$$Ln(F_o) = -rc$$

Traditionally, it is common to plot the data in semi-log plot and interpolate the number of seeding cells leading to F_0=ln(−1)≈0.379, as this corresponds to the number of seeding cells that contains on average 1 cell. Bonefoix et al. *(7)* have provided another more rigorous way analyzing the LDA data using a generalized linear model, including a simple way to check whether or not the single-hit Poisson model fits well the data. We made these options available on-line, as illustrated in the text.

5. In our settings, this method of data analysis was compatible with the sensitivity of our detection system (~1,000 cpm). Variation can be made, if more sensitive counters are used.
6. Often LDA data do not follow single-hit Poisson model, since the assumption that a single cell is sufficient to give a measurable response does not hold. For example, the TCR-dependent T cell proliferative responses in vitro are typically dependent on endogenous IL-2 produced by the T cells themselves. Therefore one must seed the microcultures with a critical number of responder cells producing IL-2, whose cooperation is necessary to kick-off a detectable response. These situations fit the multi-hit Poisson model:

$$F_o = e^{-rc} \sum_{i=1}^{m} \frac{r^{i-1}}{(i-1)!}$$

where r is the critical number of responder cells one needs to score a proliferative response. An algorithm to estimate the fraction of cells under these conditions is implemented in the online tool that we made publicly available.

Acknowledgments

This work was supported by the 3R Research Foundation Switzerland grant Nr. 92/04 (to E.P.).

References

1. Givan AL, Fisher JL, Waugh M, Ernstoff MS, Wallace PK. A flow cytometric method to estimate the precursor frequencies of cells proliferating in response to specific antigens. J Immunol Methods 1999;230:99–112.
2. Pittet MJ, Valmori D, Dunbar PR, et al. High frequencies of naive Melan-A/MART-1-specific CD8(+) T cells in a large proportion of human histocompatibility leukocyte antigen (HLA)-A2 individuals. J Exp Med 1999;190:705–15.
3. Lefkovits I, Waldmann H. Limiting Dilution Analysis of Cells in the Immune System. London: Cambridge University Press, 1979
4. Ghielmetti M, Zwicker M, Ghielmetti T, Simon MM, Villiger PM, Padovan E. Synthetic bacterial lipopeptide analogs facilitate naive CD4+ T cell differentiation and enhance antigen-specific HLA-II-restricted responses. Eur J Immunol 2005;35:2434–42.
5. Reschner A, Moretta A, Landmann R, Heberer M, Spagnoli GC, Padovan E. The ester-bonded palmitoyl side chains of Pam3CysSerLys4 lipopeptide account for its powerful adjuvanticity to HLA class I-restricted CD8+ T lymphocytes. Eur J Immunol 2003;33:2044–52.
6. Alexander J, del Guercio MF, Maewal A, et al. Linear PADRE T helper epitope and carbohydrate B cell epitope conjugates induce specific high titer IgG antibody responses. J Immunol 2000;164:1625–33.
7. Bonnefoix T, Bonnefoix P, Callanan M, Verdiel P, Sotto JJ. Graphical representation of a generalized linear model-based statistical test estimating the fit of the single-hit Poisson model to limiting dilution assays. .J Immunol 2001;167:5725–30.

Chapter 8

T Cell Epitope-Mapping by Cytokine Gene Expression Assay

Maurizio Provenzano and Giulio C. Spagnoli

Abstract

The following method describes the identification of candidate immunogenic peptides through their ability to recall an immune T-cell activation from peripheral blood mononuclear cells (PBMCs) of individuals with defined HLA–peptide restrictions that have been previously exposed to the antigen. Isolated PBMCs are plated out at a concentration of 1×10^6 cells/ml in a 200 μl medium and incubated overnight to reduce cytokine gene expression due to cell manipulation. After starving, cells are either directly stimulated with individual peptides or not stimulated and incubated from 3 to 12 h according to experimental conditions. Quantitative real-time PCR (qrt-PCR) is performed on reverse-transcribed complementary DNA (cDNA) from total RNA that is isolated from peptide-cultured PBMCs. To perform high quality qrt-PCR, primers and probes are designed to span exon–intron junctions in order to prevent amplification of genomic DNA and to produce amplicons <150 base pairs (bp). Real-time monitoring of fluorescent emission from the cleavage of sequence specific probe by the nuclease activity of *Taq* polymerase (TaqMan method) defines threshold cycles during exponential phases of amplification. Standard curves of copy numbers of the gene of interest are normalized using as reference copy numbers of control genes.

Key words: Quantitative real-time PCR, T cell, epitope mapping.

1. Introduction

This method is based on the ability of peptides to recall an immune T-cell activation. Levels of cytokine gene expression in response to ex vivo antigen recall are used to detect epitope-specific HLA-restricted T cells. The main rationale for using ex vivo techniques is based on the notion that PBMCs that bear HLA determinants relevant to the HLA–peptide restriction under investigation carry memory T cells that have been exposed in vivo to the antigen of interest, either naturally upon viral infection

Gennaro De Libero (ed.), *T Cell Protocols: Second Edition, vol. 514*

DOI 10.1007/978-1-60327-527-9_8 Springerprotocols.com

(1) or following specific vaccination procedures *(2)*. Thus, this sensitive assay might be able to identify these responses without in vitro cell expansion. The stronger the immune response induced in vivo against epitopes from the selected antigen, the easier and more specific should be the detection of memory T-cell activities. The magnitude of cytokine gene transcription upon peptide stimulation, as analyzed by qrt-PCR in serologically positive donors, represents an accurate index of the biological events following T-cell reactivation triggered by specific MHC–peptide complexes *(3)*. The sensitivity and specificity of this method are decisive in unravelling the fine relationship between MHC–peptide associations and specific immune responses, enabling the characterization of more than one HLA restriction for one individual epitope.

2. Materials

1. Ficoll–Hypaque density gradient
2. Phosphate Buffered Saline (PBS)
3. RPMI-1640 cell culture medium supplemented with 100 μg/mL Kanamycin, 10 mM Hepes, 1 mM sodium pyruvate, 1 mM Glutamax, non-essential amino acids, and 5% human serum (thereafter referred to as complete medium).
4. RNeasy Mini Kit (QIAGEN)
5. β-Mercaptoethanol (β-ME), 14.3 M
6. Ethanol (96–100%)
7. DEPC water (0.1% diethylpyrocarbonate-treated water)
8. 1 × TE solution: 1 mM Tris–HCl, 0.1 mM EDTA, pH 8.0
9. Oligo dT primers (0.5 μg/μl)
10. RT buffer 10 ×
11. $MgCl_2$ (25 mmol)
12. dNTP 10 mmol
13. DTT (1 M)
14. Superscript RT II (50 U/μl) (Invitrogen)
15. RNase
16. HotStart mix (QIAGEN)
17. Forward primers
18. Reverse primers
19. Fluorogenic probes
20. 1 × TaqMan Universal Master Mix® (Applied Biosystems, AB)

3. Methods

The methods described below outline: (1) Cell collection and culture and peptide handling, (2) RNA extraction and complementary DNA (cDNA) synthesis, (3) design of primers and probes for quantitative rt-PCR (qrt-PCR) and their reconstitution, 4) quantitative rt-PCR performance.

3.1. Cells and Peptides

Collection, culture and handling of the cells and peptide design and reconstitution are described in **Sects. 3.1.1–3.1.3**.

3.1.1. Ficoll–Hypaque Density Gradient

1. Whole blood from either healthy donors or patients is collected in 7 ml heparinized tubes (Sarstedt) by venipuncture.
2. Transfer 10 ml of whole blood into a 50 ml Falcon tube and add 30 ml PBS to reach a total of 40 ml (**Fig. 8.1a**).
3. Mix well by shaking and then add 10 ml Ficoll–Hypaque by introducing a 10 ml pipette containing Ficoll fluid (**Fig. 8.1b and c**).
4. Flow Ficoll fluid slowly from the bottom of the tube (**Fig. 8.1c**).
5. Centrifuge at 1,950 rpm for 20 min at 20°C without brake.
6. Discard by aspirating the fluid overlaying the gradient interface (serum) leaving at least 5 ml of it over the leukocyte ring (**Fig. 8.1d**).
7. Aspirate the leukocyte ring with a 10 ml pipette carefully avoiding collecting the underlaying Ficoll fluid (toxic for cells).
8. Transfer the cells in a new 50 ml Falcon tube and fill it up to 50 ml with PBS. Mix well.

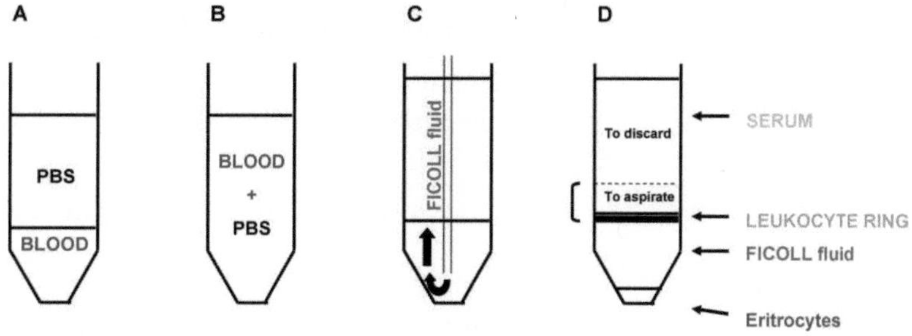

Fig. 8.1. Ficoll–Hypaque density gradient.
(**a**) Transfer 10 ml of whole blood into a 50 ml Falcon tube and add 30 ml PBS to reach 40 ml total. (**b**) Mix well by shaking. (**c**) Introduce 10 ml pipette containing Ficoll fluid and flow fluid slowly from the bottom of the tube. (**d**) Centrifuge and aspirate the leukocyte ring and the 5 ml fluid (serum) overlaying the gradient interface.

9. Centrifuge at 1,250 rpm for 15 min (long-wash) at 20°C, discard the supernatant and re-suspend the pellet in fresh PBS (50 ml total).
10. Repeat the washing at 1,250 rpm for 10 min (short wash) at 20°C.
11. Re-suspend cells (PBMCs) in RPMI complete medium (*see* **Section 2**), count them and adjust them at 1×10^6cells/ml.

3.1.2. Peptide Synthesis, Reconstitution, and Storage

Peptides in powder are reconstituted in a DMSO solution ranging from 20 to 100% depending on the solubility of the peptides. Reconstituted peptides are generally stocked in 1.5 ml cryo-vials and stored at +4°C during their use or stored at –20°C for long term storage. Peptide-stocks at 10 mg/ml concentrations are usually preferred.

3.1.3. PBMCs Plating, Culturing, and Harvesting

1. PBMCs re-suspended in RPMI complete medium at the final concentration of 1×10^6cells/ml are now plated out in 96-U bottom wells plate.
2. A total amount of 200 μl RPMI medium (2×10^5 cells total) is plated in each well.
3. The number of total loaded wells will depend on the sample condition (number of peptides) you wish to analyze plus eventual positive controls (at least one) and the negative control (cells left unstimulated) (**Fig. 8.2a**).
4. After plating, cells are rested for 18 h (normally over-night resting) in incubator (37°C, 5% CO_2, 90% humidity) to minimize cytokine gene expression due to cell manipulation.
5. After resting, cells are directly stimulated with each of the candidate peptide (plus control) at the final concentration of 1 μM peptide or left unstimulated (DMSO solution only; negative control) (**Fig. 8.2b**).
6. Cells are then re-incubated (37°C, 5% CO_2, 90% humidity) for an extra 3-h period (*see* **Note 1**).
7. Cell harvesting takes place afterwards by transferring cells from wells into 2 ml collection tube and centrifuging cells for 5 min at 1,500 rpm at room temperature (**Fig. 8.2c**).
8. Remove supernatant by aspirating carefully (*see* **Note 2**).

3.2. Genomic RNA and DNA

Total RNA extraction for each sample is performed following the RNeasy Mini Kit (QIAGEN) instructions (*see* **Note 3**): (a) disruption of the cells, (b) washing of the cell lysate, (c) elution of total RNA from disrupted cells. Up to 1 μg of total RNA is synthesized into cDNA through: (a) RNA-oligo (dT) primers pre-heating, b) RT-PCR mixture holding, (c) cDNA synthesis by reverse transcription, (d) RNA clean up.

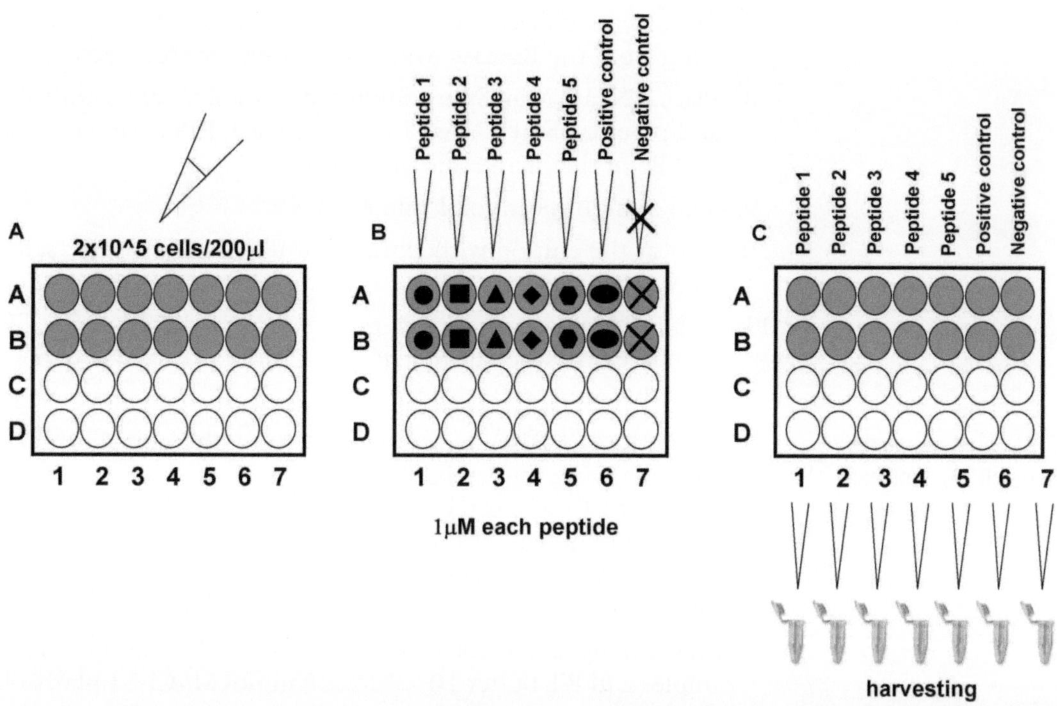

Fig. 8.2. (**a**) Plate out PBMCs in 96-U bottom wells plate at the final amount of 2×10^5cells total (200 μl RPMI medium). Rest plated cells over-night in incubator (37°C, 5% CO_2, 90% humidity) to minimize cytokine gene expression. (**b**) After starving, add directly in the well the candidate peptide at the final concentration of 1 μM. Re-incubate cells for 3 h. (**c**) Harvest cells transferring them in collection tubes.

3.2.1. Total RNA Extraction

1. Add 350 μl of RTL complete lysis buffer (10 μl β-ME per 1 ml RTL buffer; prepare solution before use) in the 2 ml collection tube containing pelleted cells (*see* Section 3.1.3. *and* **Note 4**).
2. Disrupt cells by gently vortexing and then add 350 μl 70% ethanol by mixing.
3. Transfer the lysate into the RNeasy Mini Spin column (maximum loading volume 700 μl) and spin at 16,000 rpm (thereafter referred to as high speed) for 15 s at +4°C. RNA is retained in the white membrane of the RNeasy Mini Spin columns.
4. Discard the flow-through, reuse the collector tube and wash membrane one time with 700 μl RW1 washing buffer (high speed, 15 s at +4°C).
5. Change the collector tube and wash membrane with 500 μl RPI buffer containing four volumes of 96–100% ethanol at high speed for 15 s at +4°C.
6. Discard the flow-through, reuse the collector tube and repeat this wash for 2 min.

7. Discard the collector tube and leave residual ethanol evaporating from the RNeasy Mini Spin column for few seconds.
8. Place RNeasy Mini Spin columns in a new 2 ml collector tube and pipette 25 μl RNAse free water (or DEPC water) in the middle of the membrane of the tube.
9. Spin at high speed for 1 min at +4°C.
10. Repeat the elution with extra 25 μl RNAse free water (same spinning as above).
11. Put RNA eluate on ice, mark it and immediately use it for cDNA synthesis or store it at –70°C until use (*see* **Note 5**).

3.2.2. Complementary DNA (cDNA) Synthesis

Each sample of cDNA is synthesized from the corresponding RNA by following four steps:

1. First, up to 1 μg of total RNA is incubated at 70°C for 10 min with 1 μl oligo (dT) primers (0.5 μg/μl) in a final volume of 12 μl (RNA volume is defined depending on its concentration; adjust with DEPC water to reach the 12 μl final volume).
2. After cooling at 4°C for 1 min, add 7 μl of RT-PCR mix to the sample (2 μl RT buffer 10 × , 2 μl 25 mmol $MgCl_2$, 1 μl dNTP 10 mmol, 2 μl 1 M DTT) and hold the mix at 42°C for 5 min.
3. Then, add 1 μl Superscript RT II (50 U/μl) to each sample and synthesize cDNA at 42°C for 50 min and at 70°C for 15 min.
4. Clean up for possible RNA contamination is performed by incubating samples with 1 μl RNase at 37°C for 20 min.
5. Synthesized cDNA should be stored at +4°C (or –20°C for a longer storage)

3.3. Quantitative Real-Time PCR Primers and Probe

The paragraph includes: (a) Probe labeling for TaqMan method, (b) Primers and probe design, (c) Primers and probe reconstitution.

3.3.1. Probe Labeling for TaqMan Method

The concept of real-time PCR relies of the detection of PCR products as they accumulate. Current qrt-PCR systems are based on sets of primers and fluorogenic probes. The fluorescence signal increases with each PCR amplification cycle. The PCR cycle in which fluorescence reaches a threshold value of ten times the standard deviation of baseline fluorescence emission is called the cycle threshold (C_t) and it is inversely proportional to the starting amount of target genetic material *(4)*. TaqMan Method indicates a signal generation throught the degradation of a fluorogenic probe by employing the 5'–3' nuclease activity of *Taq* DNA polymerase *(5)*. TaqMan probes are labeled at the 5'-end with the reporter dye molecule FAM (6-carboxyfluorescein; emission λ_{max}=518 nm) and at the 3'-end with the fluorescent quencher dye molecule TAMRA (6-carboxytetramethylrhodamine; emission λ_{max}=582 nm) (*see* **Note 6**).

3.3.2. Primers and Probe Design

Cytokine gene primers and probe for qrt-PCR (TaqMan method) are designed by Primer Express software but some guidelines are necessary to reach high quality qrt-PCR performance *(2, 6)*.

1. Primers should be between 15 and 20 nucleotides long and not overlapping with probe complementary sequences in order to cover an amplicon with a length $\leq$ 150 base pairs (bp). Give prevalence of better probes over primers.
2. Primers melting temperature (T_m) should be 58–60°C and probes T_m should be 68–70°C.
3. Primers and probes guanosin/cytosin (G/C) content should be between 20 and 80% (avoiding G to be on the 5'end and having C content higher than G content).
4. To avoid hairpins or loops within primers and probes, the five nucleotides at 3'end of each primer should have no more than two Gs or Cs.
5. Avoid also runs of identical nucleotides within primers.

3.3.3. Primers and Probes Reconstitution

Primers are usually shipped in a lyophilized state (unless differently requested) while probes are more often shipped in 1 × TE solution. The best performance for cytokine gene expression using qrt-PCR assay is achieved by using primers and probes at specific final concentration. In our experience, optimized concentration are: 400 nM for primers (*see* **Note** 7) and 200 nM for probes *(1, 7)*. It is thus suggested to prepare both primers and probes working stock solutions in DEPC water (or 1 × TE) at 20 μM for each primer and 12.5 μM for each probe (*see* **Sects. 3.4.1 and 3.4.4**).

3.4. Performing Quantitative Real-Time PCR

This paragraph reports indications to precisely perform qrt-PCR for cytokine gene expression analysis: (a) cDNA amplification and quantification (b) tenfold serial dilutions of known gene concentrations for standard curve generation (c) standard curve validation for absolute quantification analysis (*see* **Note 8**), (d) TaqMan Master Mix preparation and thermal cycler parameters definition for both standard and test cDNA qrt-PCR reactions.

3.4.1. cDNA Amplification and Quantification for Standard Curve Generation

Complementary DNA from a positive control template is generated by reverse transcription using a technique identical with the one used for the preparation of test cDNA (**Section 3.2.2**). Cytokine genes are amplified by means of regular PCR using the same primers designed for qrt-PCR (**Section 3.3.2**).

1. PCR mix: 25 μl HotStart mix (QIAGEN), 1 μl forward primer at 20 μM (400 nM), 1 μl reverse primer at 20 μM (400 nM), 1 μl cDNA, 22 μl DEPC water; final volume 50 μl.
2. PCR parameters:
   - 10 min at 95°C
   - 30 s at 95°C, 30 s at 60°C and 2.5 min at 72°C (40 cycles)
   - 5 min at 72°C

3. Quantify amplified cDNA (double strand) in μg/ml by spectrophotometry (OD=260).

3.4.2. Amplified Gene Serial Dilutions for Standard Curve

In order to create a curve by tenfold standard dilutions at known gene concentration, the number of copies of amplified cDNA (standard cDNA) from a positive control template has to be calculated. To perform it you should know:

1. The concentration of your amplified cDNA (OD260 reading usually in μg/ml)
2. The molecular weight (MW) of your amplicon
3. The Avogadro number (number of molecules in 1 mole = 6.023×10^{23})

Example for calculation of IFNγ molecules after DNA amplification:

1. OD260 = 461 μg/ml → 4.61 μg/10 μl (4.61 μg → 4.61×10^{-6} g).

   This is your starting concentration for serial dilutions (see below).
2. Molecular weight (MW) of IFNγ amplicon = 36,492 g/M
3. $4{,}61 \times 10^{-6}$ g/36,492 g/M → 1.26×10^{-10} M
4. 1.26×10^{-10} M * 6.023×10^{23} → 7.61×10^{13} molecules
5. 7.61×10^{13} is equivalent (~) to 1×10^{14} molecules

   You have approximately 10^{14} molecules in 10 μl of your solution.
6. Make tenfold serial dilutions adding 10 μl of your starting solution (10^{14} molecules) to 90 μl DEPC water to reach the final volume of 100 μl. Repeat dilutions as below illustrated (10 μ sample + 90 μl DEPC water) and stop with the last solution of 10^{3} molecules (**Fig. 8.3**).
7. Use cDNA serial dilutions for qrt-PCR test or store them at +4°C (or –20°C for a longer storage).

3.4.3. Quantitative rt-PCR Reactions

Quantitative rt-PCR reactions for both standard and test cDNA are conducted as reported below. To test your standard for cytokine gene expression, use dilutions from 10^{8} to 10^{3}.

1. Prepare a TaqMan Mix: 12.5 μl of 1 × TaqMan Universal Master Mix®, 0.5 μl forward primer at 20 μM (400 nM), 0.5 μl reverse primer at 20 μM (400 nM), 0.4 μl probe at 12.5 μM (200 nM), 10.1 μl DEPC water; final volume 24 μl.
2. Load 96-well optical reaction plate (Applied Biosystems) with 24 μl TaqMan Mix in triplicate for each sample you wish to run. Avoid bubbles.
3. Add cDNA to each of the three wells related to the sample you wish to run.
4. Close tightly the plate.

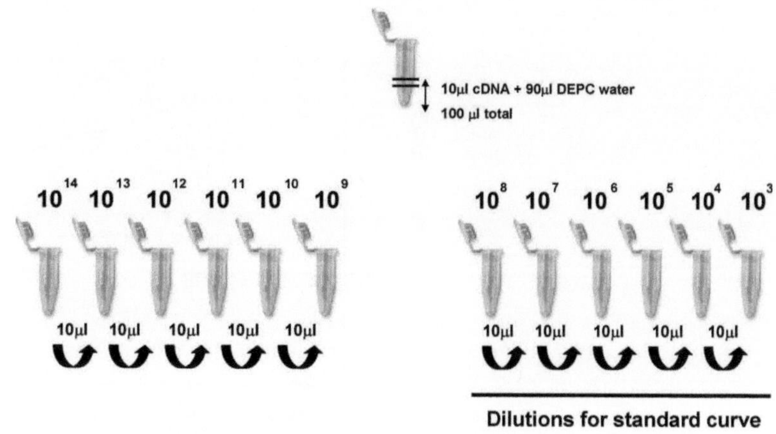

Fig. 8.3. Serial dilutions preparation.
Make tenfold serial dilutions adding 10 μl of your starting solution (i.e., 10^{14} molecules) to 90 μl DEPC water to reach the final volume of 100 μl. Repeat dilutions as above illustrated (10 μl sample + 90 μl DEPC water) and stop with the last solution of 10^3 molecules. To test your standard for cytokine gene expression, use dilutions from 10^8 to 10^3.

5. Centrifuge the plate at 2,000 rpm for 2 min at +4°C.
6. Thermal cycler parameters for 25 μl final volume (24 μl TaqMan Mix + 1 μl cDNA) using either an ABI prism 7700, 7900, or 7500FAST Sequence Detection System are:
   a. 2 min at 50°C
   b. 10 min at 95°C
   c. 15 s at 95°C (denaturation) and 1 min at 60°C (annealing/extension) for 45 cycles.

3.4.4. Standard Curve Validation and Gene Copy Number Extrapolation (Absolute Quantification Method)

1. Standard curve generation is necessary for the extrapolation of gene copy numbers (absolute quantification of gene copy numbers). Serial dilutions of the amplified gene (standard cDNA) at known concentration (10^8–10^3) are thus tested by qrt-PCR to determine the standard curve PCR for slope of −3.32, efficiency = 1. It indicates doubling of template after each PCR cycle). Linear regression analysis of each standard should show a high correlation ($R^2 \geq 0.99$) (**Fig. 8.4**). Extrapolation of copy number of each given gene (x) is conducted following this formula: $x = 10^{(\text{unknown sample Ct}-y\text{ intercept})/\text{slope}}$
2. Number of copies of an endogenous control gene (reference gene; i.e., β-actin) is also quantified. For each experimental sample, the value of both target and reference genes are extrapolated from the respective standard curve equation.

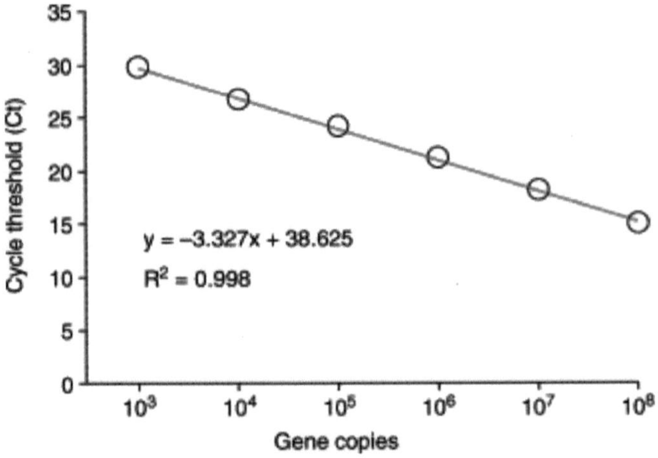

Fig. 8.4. β-Actin standard curve plots for calculation of PCR efficiency and quantitation. A tenfold serial dilution of a positive control template is used to generate the standard curve. The resulting threshold cycle (C_t) values for each input amount of template are plotted as a function of the $\log_{10}$ concentration of input amounts. The resulting slope of the line fitted to the data is used to determine the PCR efficiency. Optimal standard curves are based on PCR amplification efficiency as demonstrated by the slope of the standard curve equation. Linear regression analysis of all standard curves should show a high correlation (R^2 coefficient $\geq$0.99) to be considered suitable for gene levels quantitative analysis. The function that defines this slope is also used to calculate the amount of unknown samples. Most real-time PCR instruments have software that can automatically compute the amount of template of an unknown sample from a standard curve. However, it can be done manually by putting the observed C_t value for an unknown sample into the following formula: (observed C_t–y intercept)/slope *(1)*.

3. The target value is divided by the reference gene value to obtain normalized targets (independent from amount of starting material).

4. Notes

1. Due to differential HLA/peptide binding capacity and cytokine kinetics, incubation time could be extended to 12 h *(8)*.
2. To minimize cell handling when working with several samples, a direct centrifugation of the 96-well plate (5 min, 1,250 rpm, 20°C) after 3-h incubation could be performed and the supernatant can be vigorously flicked out.
3. For a higher quality RNA extraction, the RNA extraction Triazol method could be also chosen.
4. RLT complete lysis buffer (RTL + β-ME) can be stored at room temperature for up 1 month. RTL lysis solution (RTL complete lysis buffer + pellet) could be stored at –70°C for months.

5. Due to the instability of RNA at room temperature, all RNA extraction steps and centrifugations should be performed at 0–4°C (ice temperature). Thaw RNA out on ice when it comes from –70°C storage.
6. More reporter dyes are available to label TaqMan probes: TET™, JOE™, VIC™. By using probes labeled with different fluorochromes characterized by unique emission spectra, more genes can be analyzed simultaneously within a given sample (multiplex qrt-PCR) *(9)*.
7. Depending on the efficiency of the primers, their optimized concentration should range between 400 and 800 nM
8. As an alternative to absolute quantification method (and more specifically to skip standard curve generation), you can use the relative quantification method ($2^{-\Delta\Delta C_t}$) *(10)*
   - Step 1: C_t target gene (i.e., IFNγ) – C_t reference gene (i.e., β -actin) = ΔC_t.
   - Step 2: ΔC_t peptide stimulation – ΔC_t no stimulation = $\Delta\Delta C_t$
   - Step 3: $2^{-\Delta\Delta C_t}$

   This method is used to determine variations among gene expression (relative increase or decrease). For cytokine gene expression measurements upon peptide induction, fold increase of target genes in peptide stimulated cells as compared to baseline gene expression from unstimulated cells can be used *(11)*.

References

1. Provenzano M, Mocellin S, Bonginelli P, et al. Ex vivo screening for immunodominant viral epitopes by quantitative real time polymerase chain reaction (qRT-PCR). J Transl Med 2003;1:12.
2. Kammula US, Marincola FM, Rosenberg SA. Real-time quantitative polymerase chain reaction assessment of immune reactivity in melanoma patients after tumor peptide vaccination. J Natl Cancer Inst 2000;92:1336–44.
3. Ghei M, Stroncek DF, Provenzano M. Analysis of memory T lymphocyte activity following stimulation with overlapping HLA-A*2402, A*0101 and Cw*0402 restricted CMV pp65 peptides. J Transl Med 2005;3:23.
4. Provenzano M, Mocellin S. Complementary techniques: validation of gene expression data by quantitative real time PCR. Adv Exp Med Biol 2007;593:66–73.
5. Mocellin S, Rossi CR, Pilati P, Nitti D, Marincola FM. Quantitative real-time PCR: a powerful ally in cancer research. Trends Mol Med 2003;9:189–95.
6. Mocellin S, Provenzano M, Rossi CR, et al. Use of quantitative real-time PCR to determine immune cell density and cytokine gene profile in the tumor microenvironment. J Immunol Methods 2003;280:1–11.
7. Kammula US, Lee KH, Riker AI, et al. Functional analysis of antigen-specific T lymphocytes by serial measurement of gene expression in peripheral blood mononuclear cells and tumor specimens. J Immunol 1999; 163:6867–75.
8. Lim JB, Provenzano M, Kwon OH, et al. Identification of HLA-A33-restricted CMV pp65 epitopes as common targets for CD8(+) CMV-specific cytotoxic T lymphocytes. Exp Hematol 2006;34:296–307.
9. Wittwer CT, Herrmann MG, Gundry CN, Elenitoba-Johnson KS. Real-time multiplex PCR assays. Methods 2001;25:430–42.

10. Livak KJ, Schmittgen TD. Analysis of relative gene expression data using real-time quantitative PCR and the 2(-Delta Delta C(T)) Method. Methods 2001;25:402–8.
11. Provenzano M, Bracci L, Wyler S, et al. Characterization of highly frequent epitope-specific CD45RA+/CCR7+/- T lymphocyte responses against p53-binding domains of the human polyomavirus BK large tumor antigen in HLA-A*0201+ BKV-seropositive donors. J Transl Med 2006;4:47.

Chapter 9

Cytokine Multiplex Immunoassay: Methodology and (Clinical) Applications

Wilco de Jager, Berent Prakken, and Ger T. Rijkers

Abstract

Subsets of T cells can be distinguished on basis of their cytokine production and secretion profile.

With the critical role of cytokines in the regulation of immune and inflammatory responses, cytokines hold the promise to become the ideal biomarkers to monitor development and progression of immune-mediated diseases, study the development of new therapeutic approaches (both in vitro and in vivo) and as outcome parameters. Because of the numerous interactions in the cytokine network, the pleiotropic actions and redundancy, it will be necessary to monitor the complete spectrum of cytokines. As such, the multiplex immunoassay (MIA) is the ideal technique for that purpose. This paper reviews the critical methodological steps of this technique, including the procedures for antibody coupling to beads and matrix effects of biological fluids and buffer solutions. The power and robustness of the MIA technique is illustrated by an analysis of cytokine profiles in juvenile idiopathic arthritis.

Key words: Cytokines, multiplex immunoassay, Luminex, juvenile idiopathic arthritis.

1. Introduction

Primary activation of naive $CD4^+$ helper T lymphocytes (which are termed Th0 cells) on antigen presenting dendritic cells leads to differentiation into two major categories of T-helper cells: T-helper 1 cells (Th1) and T-helper 2 cells (Th2). Th1 and Th2 cells can be distinguished on basis of their cytokine production and secretion profile (**Fig. 9.1**). Accordingly, Th1 cells secrete the cytokines γ-interferon and interleukin-2 (IL-2), major cytokines for induction and regulation of the cellular immune response and for macrophage activation. Th2 cells produce and secrete IL-4, IL-5, and IL-13, cytokines which regulate the response of B

Gennaro De Libero (ed.), *T Cell Protocols: Second Edition, vol. 514*

DOI 10.1007/978-1-60327-527-9_9 Springerprotocols.com

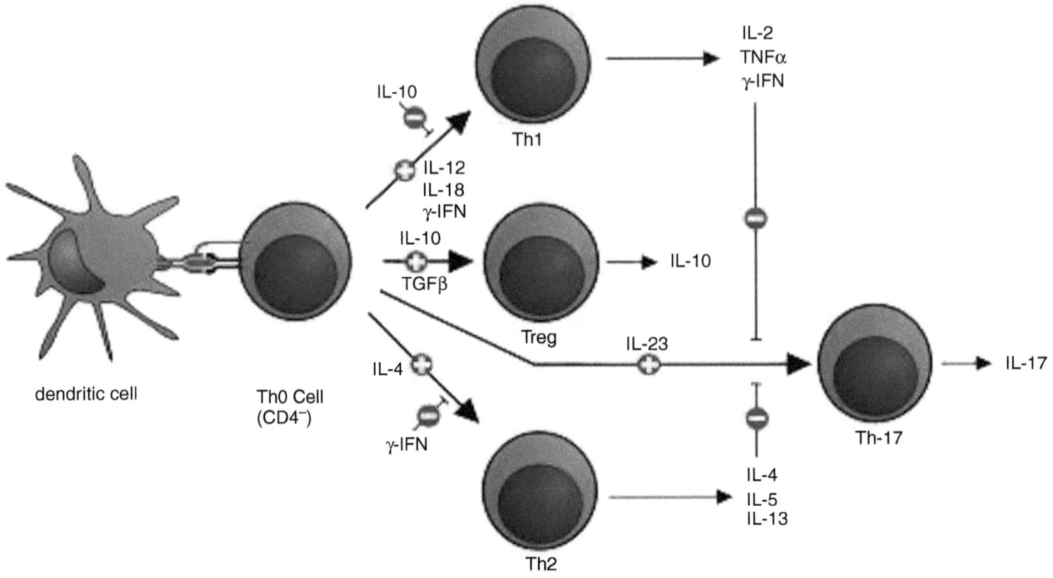

Fig. 9.1. Cytokine production pattern and regulation by subsets of CD4 T lymphocytes.

lymphocytes. Regulatory T lymphocytes (see below) produce and secrete mainly IL-10 and/or transforming growth factor β (TGF-β). Recently, a novel Th subset has been defined, Th17 *(1, 2)*. Driven by IL-23, Th17 cells produce the pro-inflammatory cytokine IL-17 (**Fig. 9.1**).

Abnormalities in regulatory circuits as found in autoimmune and allergic diseases are reflected in the cytokine pattern in blood as well as in in vitro systems. Rheumatoid arthritis is considered to be a so-called Th1-mediated autoimmune disease. In peripheral blood of these patients, increased concentrations of typical Th1 cytokines as γ-interferon and IL-2 are found *(3)*. Also pro-inflammatory cytokines such as IL-1α, IL-1β, IL-6, and TNFα are increased as compared to other forms of arthritis and to healthy controls *(3, 4)*. Moreover, also other cytokines, such as the IL-17 mentioned above, can be increased *(4)*. Not always (or maybe even: most of the times not) is the cytokine pattern as found in blood a good reflection of the situation at the site of active inflammation, in case of RA the affected joints. In such as case analysis of cytokines in synovial fluid is indicated.

The two major subpopulations of T lymphocytes are the CD4 (helper) cells and the CD8 cells. The CD8 T lymphocyte population includes cytotoxic T lymphocytes and until the mid-1980s of the last century within this subset also suppressor T lymphocytes were defined. Research into suppressor T lymphocytes appeared to be difficult and gradually faded away. Recently there has been a modest revival of CD8 regulatory cells. These cells are capable to recognize autoantigens, presented by non-classical class 1 MHC molecules (HLA-E) on activated, autoreactive T lymphocytes, and to suppress or kill these cells *(5)*.

Within the $CD4^+$ T lymphocytes a separate category of regulatory T lymphocytes is formed by the $CD4^+CD25^+$ regulatory T cells (Tregs) *(6, 7)*. Tregs have the ability to suppress T cell responses and thus can play a pivotal role in the regulation of the immune response.

The discovery of $CD4^+CD25^+$ Tregs came from experiments in mice in which it was found that neonatal thymectomy caused a multiorgan autoimmune disease. The crucial experiments were conducted in the laboratory of Sakaguchi, where it was found that injection of $CD25^-$ T lymphocytes into nude mice leads to a systemic autoimmune disease with signs of van colitis, gastritis, autoimmune diabetes, and thyroiditis *(8)*. Addition of low numbers of $CD4^+CD25^+$ T lymphocytes to the $CD25^-$ T lymphocytes completely prevented development of the disease *(8)*. Also in humans $CD4^+CD25^+$ T lymphocytes, despite constituting only a small fraction of $CD4^+$ cells, are important regulatory cells.

With the critical role of cytokines in the regulation of immune and inflammatory responses, cytokines hold the promise to become the ideal biomarkers to monitor development and progression of immune-mediated diseases, study the development of mediation (both in vitro and in vivo) and as outcome parameters. Because of the numerous interactions in the cytokine network, the pleiotropic actions and redundancy, it will be necessary to monitor the complete spectrum of cytokines. As such, the multiplex immunoassay is the ideal technique for that purpose.

Cytokine production by T lymphocytes in principle can be measured at three levels: mRNA, intracellular production (most often expressed as the frequency of cells which produce a particular cytokine) or at the level of the actual secreted cytokines. Every assay has its pro's and con's with large variations in amount of cells, time and money required. The ideal assay once has been characterized by Vignalli in that it should require a minimal amount of cells, time and money but also should be specific, sensitive, simple (the KISS principle; keep it simple, stupid) and reliable, or "robust" as today's alternative for reliability *(9)*. The multiplex immunoassay for measurement of cytokines fulfills most of these criteria for the perfect assay as will be indicated below. We advocate the acronym MIA for this multiplex immunoassay. MIA would be a logical extension of a series of laboratory techniques that began with RIA (for radio immunoassay) and was continued with ELISA (enzyme-linked immunosorbent assay).

2. Materials

2.1. Covalent Coupling of the Capture Antibodies

1. Antibodies and recombinant proteins (**Table 9.1**)
2. *N*-hydroxy-sulfosuccinimide (Sulfo-NHS; Pierce Rockford, USA)
3. 1-Ethyl-3-(3-dimethylaminopropyl)-carbodiimide hydrochloride (EDC; Pierce, Rockford, USA)
4. xMAP COOH microspheres (Biorad Laboratories, Hercules USA) *Light sensitive*
5. Vivaspin 500, 10,000 Kd (Vivascience, Lincoln, UK)
6. Phosphate-buffered saline 0.1 M, pH 7.2 (PBS)
7. PBS, 0.05% Tween 20
8. PBS, 0.05% Tween 20, 1% bovine serum albumin
9. PBS, 1% BSA, 0.1% sodium azide
10. Goat anti mouse IgG fluorescein isothiocyanate (FITC; BD Bioscience, San Diego, USA)

Table 9.1
Reagents used for the multiplex immunoassay

Mediator	Protein	Capture antibody			Detection antibody		
	Source	Clone	Species	Source	Clone	Species	Source
IL-2	BD	5344.111	Mouse	BD	B33-2	Mouse	BD
IL-4	RD	8D4-8	Mouse	EB	MP4-25D2	Rat	EB
IL-5	BS	JES1-39D10	Rat	BS	JES1-5A10	Rat	BD
IL-10	BD	JES3-19F1	Rat	BD	JES3-12G8	Rat	BD
IL-13	NIBSC	CLB/IL13-1	Mouse	SQ	CLB/13-2	Mouse	SQ
IL-17	RD	41809.111	Mouse	RD	Polyclonal	Goat	RD
IL-23	EB	eBio473P19	Rat	EB	C8.6	Mouse	EB
TNFα	SM	MAb1	Mouse	BD	MAb11	Mouse	BD
IFNγ	BD	NIB42	Mouse	BD	4S.B3	Mouse	BD

The reagents listed were obtained from the source indicated by the following abbreviations; BD = BD Biosciences (San Diego, CA, USA), BS = Bioscource/Invitrogen (Nivelles, Belgium), EB = eBioscience (San Diego, CA, USA), NIBSC = National Institute for Biological Standards and Control (Potters Bar, UK), RD = R&D Systems (Abington, UK), SM = Strathmann Biotec (Hamburg, Germany), SQ = Sanquin (Amsterdam, The Netherlands).

2.2. Multiplex Immunoassay

1. Coupled microspheres, biotinylated antibodies and recombinat proteins (**Table 9.1**)
2. Streptavidin R-phycoerythrin (SA-PE; BD Biosciences, San Diego, USA)
3. PDVF filterplates (96 well) (Millipore, Billerica, USA)
4. High Performance ELISA Buffer (HPE, Sanquin, Amsterdam, The Netherlands)
5. Serum diluent (R&D Systems, Abington, UK)
6. Bio-plex Suspension Array System Luminex (xMAP-based platform, Biorad Laboratories, Hercules, USA)

2.3. Matrix Effect

1. PBS, 5% FCS (Gibco / Invitogen, Gaithersburg, USA)
2. PBS, 5% human AB serum (Sanquin Bloodbank, Amsterdam, The Netherlands)
3. AIM-V (Gibco/ Invitrogen, Gaithersburg, USA) supplemented with 5% FCS
4. RPMI1640 (Gibco/ Invitrogen, Gaithersburg, USA) supplemented with 5% FCS
5. IMDM (Gibco/ Invitrogen, Gaithersburg, USA) supplemented with 5% FCS
6. D-MEM (Gibco/ Invitrogen, Gaithersburg, USA) supplemented with 5% FCS
7. Various recombinant proteins (5 ng/ml, **Table 9.1**)

2.4. Cell Culture for Cytokine Analysis

1. Ficoll (1.077 g/cm^3; Amersham Pharmacia Biotech AB, Uppsala, Sweden)
2. RPMI1-640 supplemented with 5% FCS
3. CD3 (clone OKT3, eBioscience)
4. CD28 (clone CD28.2, eBioscience)
5. Tetanus toxoid (TT; National Institute of Public Health and the Environment [RIVM], Bilthoven, The Netherlands)

2.5. Assay Performance

1. Ficoll (1.077 g/cm^3; Amersham Pharmacia Biotech AB, Uppsala, Sweden)
2. RPMI1640 supplemented with 5% FCS
3. Concanavalin A (ConA; Calbiochem, La Jolla, USA)
4. Tetanus toxoid (TT; National Institute of Public Health and the Environment [RIVM], Bilthoven, The Netherlands)
5. Phytohemagglutinin (PHA; Murex Biotech, Dartford, UK)

3. Methods

The methods described below outline (**Section 3.1**) the covalent coupling of the capture antibodies to the carboxylated microspheres, (**Section 3.2**) the multiplex immunoassay itself, (**Section 3.3**) any matrix effects that may occur, (**Section 3.4**) implementation of cytokines measurements in cell culture supernatants and (**Section 3.5**) assay performance overtime.

3.1. Covalent Coupling of Antibodies to the Microspheres

All antibody pairs used were directed against different non-competing epitopes of their respective cytokines and were purchased from different commercial sources (**Table 9.1**). If necessary, antibodies were reconstituted in phosphate-buffered saline (PBS), pH 7.4. Sodium azide (NaN3) was removed from the capture antibodies with a Vivaspin 500 concentrator with a 10,000-molecular-weight cutoff polyethersulfone membrane (Vivascience, Lincoln, UK), which was spun three times at 10,000 × *g* with PBS used as the wash matrix. All recombinant proteins except IL-13 were reconstituted in PBS, pH 7.4, containing 1% bovine serum albumin (BSA; Sigma-Aldrich, Zwijndrecht, The Netherlands) to a concentration of 10 μg/ml. IL-13 was reconstituted to a concentration of 1 μg/ml. All proteins were aliquoted and stored at –80°C. Covalent coupling of the capture antibodies to the microspheres was performed by following the procedures recommended by Luminex Corporation (Austin, USA, **Fig. 9.2A**) using *N*-hydroxy-sulfosuccinimide (Sulfo-NHS; Pierce Rockford, USA) and 1-ethyl-3-(3-dimethylaminopropyl)-carbodiimide hydrochloride (EDC; Pierce, Rockford, USA), as follows:

1. The microspheres stock solutions were dispersed in a sonification bath for 1–2 min. An aliquot of 2.5 $\times 10^6$ microspheres was resuspended in microtiter tubes (Eppendorf, Hamburg, Germany) containing 0.1 M sodium phosphate buffer, pH 6.1 (phosphate buffer), to a final volume of 80 μl.
2. This suspension was mixed until a homogeneous distribution of the microspheres was observed.
3. Solutions of Sulfo-NHS and EDC, both at 50 mg/ml, were freshly prepared in phosphate buffer, and 10 μl of each solution was sequentially added to stabilize the reaction and activate the microspheres.
4. This suspension was incubated for 10 min at room temperature and then resuspended in 250 μl of PBS containing 50 μg of antibody (*see* **Note 1**).
5. The mixture was incubated overnight in the dark with continuous shaking.

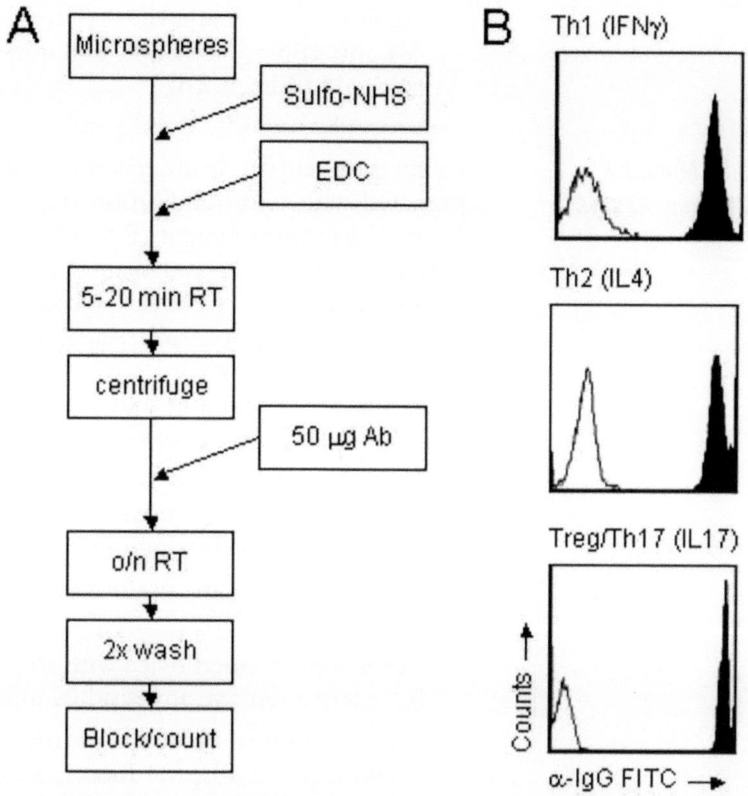

Fig. 9.2. Flow chart of coupling procedure of anti-cytokine antibodies to beads (panel **A**) and flowcytometric analysis of coupled beads (panel **B**). Open histograms, uncoupled beads, filled histograms, beads coupled with (from *top* to *bottom*) antibodies against IFNγ, IL-4, or IL-17, respectively. FITC fluorescence intensity is plotted on the *X*-axis, the relative number of beads on the *Y*-axis.

6. Microspheres were then incubated with 250 µl of PBS-0.05% Tween 20 for 4 h at RT.
7. After aspiration, the beads were blocked with 1 ml of PBS-1% BSA-0.1% sodium azide.
8. The microspheres were counted with a hemacytometer and stored at a final concentration of 10^7 microspheres per ml in the dark at 4°C.
9. Coupling efficiency of monoclonal antibodies was tested by staining 5,000 microspheres with either fluorescein isothiocyanate (FITC)-conjugated goat anti-mouse immunoglobulin G (IgG; BD Biosciences, San Diego, CA, USA) or FITC-conjugated goat anti-rat IgG antibodies (Zymed Laboratories, San Francisco, CA, USA).
10. When the microspheres are analyzed using a regular flow cytometer before and after coupling a clear positive signal is found on the coupled microspheres. Representative examples

of flow cytometric analysis of beads before and after coupling of anti-IFNγ, anti-IL-4, and anti-IL-17 antibodies are displayed in **Fig. 9.2B**.

3.2. Multiplex Cytokine Assays

Calibration curves from recombinant cytokine standards were prepared with twofold dilution steps (13 points and a blank, *see* **Note 2**) in serum diluent (R&D Systems, Abington, UK). Samples were measured, and blank values were subtracted from all readings. All assays were carried out directly in a 96-well PDVF filterplates (Millipore, Billerica, USA) at room temperature and protected from light.

1. A mixture containing 5,000 microspheres per cytokine (total volume 10 μl/well) was incubated for 1 h.
2. Next 10 μl/well of a cocktail of biotinylated antibodies (16.5 ng of each biotinylated antibody) was added and incubated for an additional hour.
3. All incubations are performed at room temperature under continuous shaking.
4. Beads were washed twice with PBS-1% BSA-0.05% Tween 20 in order to remove any residual unbound antibody.
5. After 10 min of incubation with 100 ul Streptavidin R-phycoerythrin (SA-PE; BD Biosciences, San Diego, USA) in PBS (50 ng/well) and washing with PBS-1% BSA-0.05% Tween 20, the fluorescence intensity of the beads was measured in a final volume of 100 μl of HPE buffer (High Performance ELISA buffer; Sanquin, Amsterdam, The Netherlands).

Measurements and data analysis of all assays were performed with the BioPlex system in combination with the Bio-Plex Manager software (version 4.1) (Bio-plex Suspension Array system Luminex (xMAP based platform, Biorad Laboratories, Hercules, USA). This system distinguishes and classifies each microsphere by making use of the specific fluorescence dyes that are internalized in the microspheres. The SA-PE reporter signal does not interfere with the classification signals. A flow chart of the complete procedure is given in **Fig. 9.3**.

Accuracy of a quantitative cytokine immunoassay depends, amongst others, on the quality of the standard curves. These curves consist of a series (ranging from 2 to 13) of known amounts of natural or recombinant cytokine. In general, the cytokine concentrations are plotted versus absorbance or fluorescence intensity and, subsequently, this plot is used for a mathematically approach for curve fitting. An example of standard curves for Th1, Th2 and Treg/Th 17 subset are shown by **Fig.** 9.4. The type of curve fitting has a direct effect on the accuracy of the results. The most common mathematical approach (often used in ELISA) is the linear regression model. When the biological range of cytokines

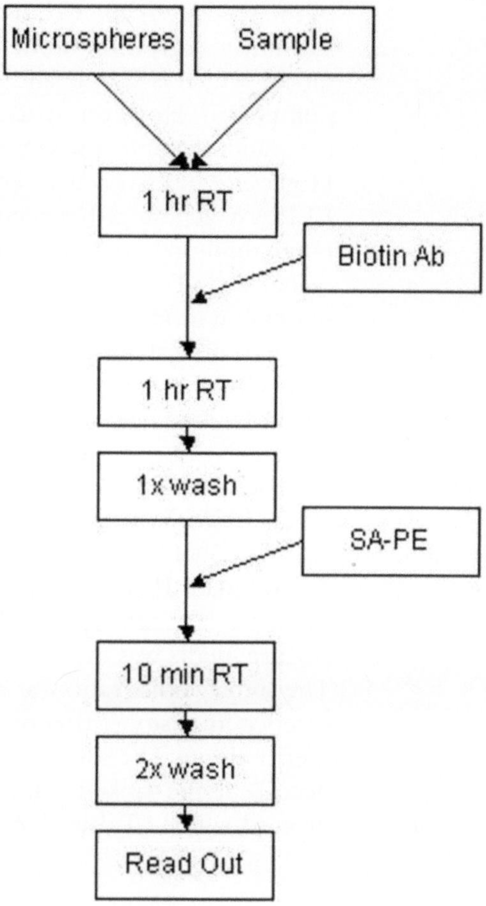

Fig. 9.3. Flow chart of cytokine multiplex immunoassay.

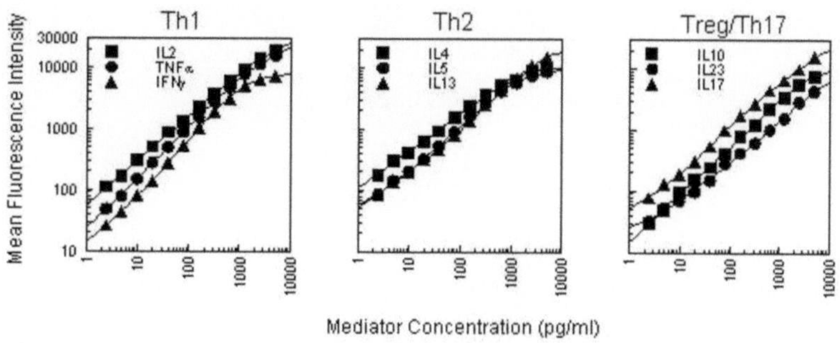

Fig. 9.4. Standard calibration curves for Th1 (*left panel*), Th2 (*middle panel*) and regulatory T cell / Th17 cytokines (*right panel*). Recombinant or purified cytokines, in a concentration range of 2.4–5,000 pg/ml were incubated with relevant beads and detection antibodies and analyzed by MIA. Data are derived from a single, representative 30-plex experiments. For clarity, IL-2, TNFα, and IFNγ curves are grouped (as Th1 cytokines), as are IL-4, IL-5, and IL-13 (as Th2 cytokines) and IL-10, IL-23, and IL-17 (Treg/Th17).

of the specimens are within the linear range of the standard curve, this linear regression model is appropriate. Linear regression is a model which is simple and can be based on as few as three data points *(10)*. However, when the data are outside the linear part of the standard curve, the samples have to be diluted and reanalyzed. However, bead-based assays have a broader dynamic range as ELISA which range between 3 and 4 log scale. A broader dynamic range minimizes the number of samples that have to be rerun and reanalyzed. Though, with a broad range standard curve, a linear regression model will not be sufficient to create a large linear part of the standard curves. Therefore a non-linear regression model with either a four-parameter (4PL) or a five-parameter (5PL) approach is to be preferred *(11)*.

3.3. Matrix Effects

To asses interference of different matrices, widely used cell culture media (AIM-V, RPMI1640, IMDM, an D-MEM, all supplemented with 5% FCS) were spiked with a fixed amount (5 ng/ml) of recombinant IL-2, IL-4, IL-10, TNFα, and IFNγ (as indicated by the material **Section 2.3**). Next both the spiked sample and the culture media was measured in triplicates over a period of 3 days. The signal/noise ratio was calculated by dividing the mean fluorescence intensity (MFI) of the spiked sample by the MFI of the media alone. The ideal assay will be formed when a high signal occurs in the spiked sample in combination with a low-background signal (**Table 9.2**). In our example, the serum diluent

Table 9.2
Signal noise ratios obtained in various tissue culture media

Medium	Signal/noise ratio
5% Human AB in PBS	206 ± 28
5% FCS in PBS	113 ± 11
AIM-V	365 ± 45
D-MEM	149 ± 56
IMDM	211 ± 21
RPMI1640	293 ± 34
Serum diluent	943 ± 41

Different culture media and buffers were spiked with 5 ng/ml IL-2, IL-4, IL-10, TNFα, and IFNγ. Next spiked samples and the culture media were measured in triplicates over a period of 3 days. The signal/noise ratio was calculated by dividing the mean fluorescence intensity (MFI) of the spiked sample by the MFI of the media alone.

yielded the best result. However, since this is a serum diluent developed for ELISA purposes, it is not very suitable for analysis of cell culture supernatants. Because of the rather large matrix effects that can be found, we strongly recommend to construct the calibration curves always in the same tissue culture medium as used for the actual cell cultures themselves.

3.4. Analysis of In Vitro Cytokine Production Patterns

Application of the MIA technique on supernatants of in vitro activated T cells allows semi-high throughput analysis of cytokine responses, again with only small sample sizes required. As an example of the power of the technique the data of our analysis of immunodysregulation in an autoimmune disease of childhood will be presented.

1. Heparinized blood samples were collected from three patients with juvenile idiopathic arthritis who visited the outpatient clinic of at our hospital, as well as from three healthy control donors. Informed consent was obtained either from parents or directly from the patients when they were older than 12 years.
2. Peripheral blood mononuclear cell (PBMC) were isolated by Ficoll density gradient centrifugation (Ficoll 1.077 g/cm^3; Amersham Pharmacia Biotech AB, Uppsala, Sweden).
3. All cultures were performed in RPMI 1640 tissue culture medium supplemented 5% heat-inactivated FCS.
4. Cells were cultured at 2×10^5 cells per well and incubated for 24 h at 37°C in either RPMI medium alone, with 1.5 μg/ml of tetanus toxoid (TT; National Institute of Public Health and the Environment [RIVM], Bilthoven, The Netherlands), or with plate bound CD3/CD28.
5. To that end round-bottomed microtiter plates were coated with 1 μg CD3 and 1.5 μg CD28; clones OKT2 and CD28.2, respectively, both from eBioscience) per well during 1 h at 37°C.
6. At the end of the culture period, cell-free supernatants were collected and stored at −80°C until analysis.
7. Cytokine analysis was performed using the multiplex immunoassay as described in **Section 3.2**.

The mathematic analysis of the obtained cytokine response can be performed using various different statistical packages (*see* **Note 3**). However, using a standard statistical approach, comparison can only be made on single cytokine level, loosing many of the advantages of a multiplex assay. Therefore, as shown in **Fig. 9.5**, a heat map can be constructed using raw data of each individual showing all cytokines. Using this approach, it is directly apparent if cytokines are up or down regulated in various individuals and to detect groups of cytokines which cluster together,

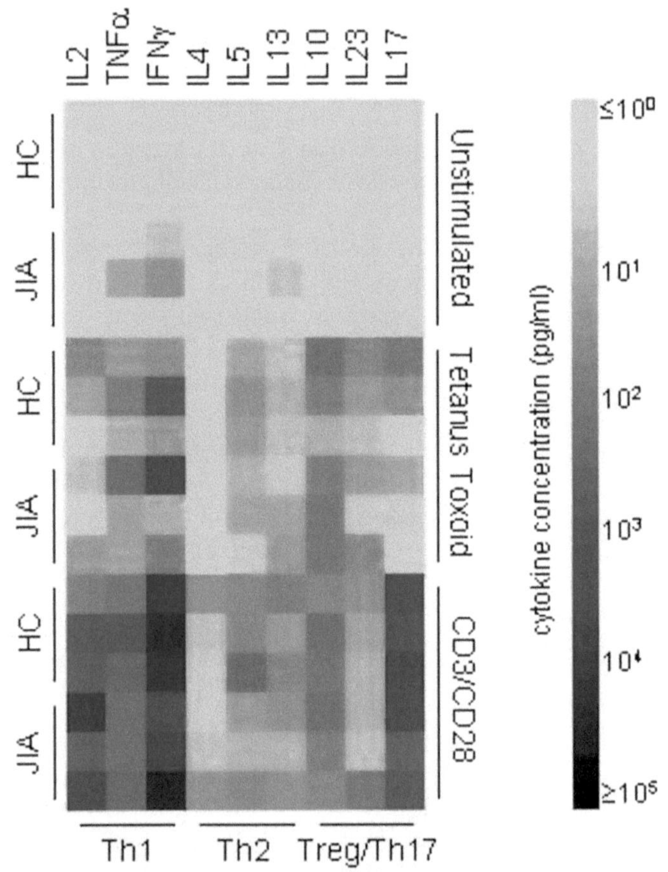

Fig. 9.5. Multiplex immunoassay of T cell cytokines in supernatants of peripheral blood mononuclear cells of healthy controls (HC) or patients with juvenile idiopathic arthritis (JIA) which were cultured with tetanus toxoid, with a combination of CD3 and CD28 antibodies or were left unstimulated. The individual cytokine concentration of an individual culture is indicated according to the pseudocolor scale on the right hand side of the figure.

without any specific statistical approach. This is very useful since the MIA will generate large amounts of data in a short period of time. Here we show after 24 h using TT or CD3/CD28 significant cytokine released into the culture supernatant. The data in **Fig. 9.6** illustrate that during a recall response to tetanus toxoid, several cytokines are upregulated, including IL-2, TNF-α, IFN-γ, IL-5, IL-13, IL-10, IL-23, and IL-17. The observed cytokine levels vary between individuals, probably depending on the vaccination status and/or in vivo exposure. Following polyclonal T cell activation with a combination of CD3 and CD28 antibodies, cultures of PBMCs of both JIA patients as well as healthy controls generated higher levels of most of the cytokines mentioned above, now also including IL-4. Next to the heat map approach for

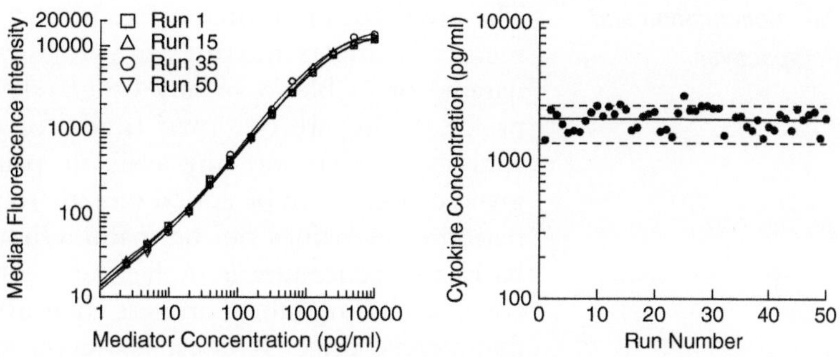

Fig. 9.6. Standard curves and intra-assay variation. *Left panel* shows the data points to be used for construction of a (12 point) calibration curve. Data of four independent experiments, run several months apart, are shown. *Right panel* shows longitudinal variation in the observed concentration of IL-2 as measured in a positive control sample (consisting of the supernatant of polyclonal activated mononuclear cells, aliquotted and stored at –80°C). The 50 runs were performed over a period of 18 months. *Horizontal drawn line* is the median value, *dotted lines* are + and –30% of the median.

statistical analysis, a cluster analysis can be performed but such an analysis requires larger datasets. At any rate, the MIA technique is capable to detect and quantify the in vitro cytokine response of T lymphocytes, activated either polyclonally or with specific recall-antigens.

3.5. Assay Performance

To asses overtime assay performance an internal control sample was run every assay.

1. The internal control was made by isolating cells as described at **Section 3.4**.
2. To induce a cytokine profile cells were cultured at 2×10^5 cells per well and incubated for 36 h at 37°C in RPMI medium supplemented with 2.5 μg/ml concanavalin A (ConA), 1.5 μg/ml of tetanus toxoid and 7 μg/ml of phytohemagglutinin (PHA).
3. At the end of the culture period, supernatants were collected and stored at −80°C until analysis.
4. After each run data was plotted and a Levy–Jennings plot was constructed (**Fig. 9.6**).
5. Recombinant cytokines that are used to construct standard curves can degrade during long time storage. This will result in a downward shift of MFI from the standard curves. Therefore it is useful to create overlays from standard curves over time (**Fig. 9.6**). Furthermore when this phenomenon occur absolute values of an internal control sample will increase overtime. Thus assay performance can simply monitored using a single control sample.

3.6. Conclusions and Perspectives

The multiplex immunoassay is a powerful technique for measurement of complex mixtures of cytokines in T cell culture supernatants or in blood or any other relevant biological fluid. We prefer to prepare our own beads for cytokine and chemokine analysis. In that way any available pair of antibodies against a given cytokine can be added directly to the panel and any combination of cytokines can be made. On a per cytokine per sample basis, this procedure is by far the most cost-effective. The total costs, excluding labor, are less than 50% of the lowest available commercial price. It does, however, require specific laboratory infrastructure and a certain turnover in order to become really cost-effective. For the more "casual" user of the technique, now several companies offer versatile and competitively priced kits for measurement of human or murine cytokines. Thus, multiplex analysis of T cell cytokines is within reach of everyone studying immunoregulation.

4. Notes

1. The water soluble Sulfo-NHS is used to modify a carboxyl group to an amine reactive NHS ester. This is accomplished by mixing the NHS with a carboxyl containing molecule and a dehydrating agent (EDC). The carboxyl group is present on the microspheres that are used for the multiplex immuno assay. EDC reacts with the carboxyl group first and forms an amine-reactive intermediate, an *O*-acylisurea. The intermediate is unstable in aqueous solutions and is therefore not useful in two-step conjugations procedures without stabilization of the intermediate using NHS *(12, 13)*. Failure to react with an amine will result in hydrolysis of the intermediate, regeneration of the carboxyl, and release of an *N*-substituted urea. This will occur with prolonged incubation times (>20 min). A side reaction is the formation of an *N*-acylurea, which is usefully restricted to carboxyl groups located in hydrophobic usually of antibodies *(14)*.
2. Calibration curves can also be made based using a higher dilution factor and can go down to as low as 6 data points, since both 4PL and 5PL regression models require at least 6 data points. However, additional data points to create a standard curve will also result in a better and more accurate fit, and thus a more accurate quantitation across a broader range of the standard curve compared *(10)*. The difference between a 4PL and a 5PL logistic curve-fitting model is an addition parameter which is added for asymmetry in a 5PL model. This additional parameter provides a better fit when standard curves are not symmetrical. The choice of logistic regression that will yield

the best fit of a given dataset is dependent on the shape of the standard curve *(11)*. When a sigmoidal curve is obtained there will no differences in outcome when samples are analyzed with either a 4PL or a 5PL regression model. However, when curves are not symmetrical due to low or high sensitivity of the antibody set used in the assay interpretation with a 5 PL regression model will yield better results *(15)*.

3. When cytokines from cell cultures are analyzed in cohorts (with or without stimulation) a large natural dynamic variation will be present. Due to this large biological variability in cytokine levels and non-Gaussian distribution it can be envisioned that a single or a few outliers in a population can scrutinize the data set when mean values are calculated. Therefore we recommend using geometric means when data is presented as well when heat maps are constructed with mean cytokine levels of a population.

References

1. Park H, Li Z, Yang XO, et al. A distinct lineage of CD4 T cells regulates tissue inflammation by producing interleukin 17. Nat Immunol 2005;6:1133–41.
2. Harrington LE, Hatton RD, Mangan PR, et al. Interleukin 17-producing CD4+ effector T cells develop via a lineage distinct from the T-helper type 1 and 2 lineages. Nat Immunol 2005;6:1123–32.
3. de Jager W, Prakken BJ, Bijlsma JW, Kuis W, Rijkers GT. Improved multiplex immunoassay performance in human plasma and synovial fluid following removal of interfering heterophilic antibodies. J Immunol Methods 2005;300:124–35.
4. de Jager W, Hoppenreijs EP, Wulffraat NM, Wedderburn LR, Kuis W, Prakken BJ. Blood and synovial fluid cytokine signatures in patients with juvenile idiopathic arthritis: a cross-sectional study. Ann Rheum Dis 2007;66:589–98.
5. Jiang H, Chess L. Regulation of immune responses by T cells. N Engl J Med 2006;354:1166–76.
6. Hori S, Takahashi T, Sakaguchi S. Control of autoimmunity by naturally arising regulatory CD4+ T cells. Adv Immunol 2003;81:331–71.
7. Sakaguchi S. Regulatory T cells: Meden Agan Immunol Rev 2006;212:5–7.
8. Sakaguchi S, Sakaguchi N, Asano M, Itoh M, Toda M. Immunologic self-tolerance maintained by activated T cells expressing IL-2 receptor alpha-chains (CD25). Breakdown of a single mechanism of self-tolerance causes various autoimmune diseases. J Immunol 1995;155:1151–64.
9. Vignali DA. Multiplexed particle-based flow cytometric assays. J Immunol Methods 2000;243:243–55.
10. Motulsky HJ, Ransnas LA. Fitting curves to data using nonlinear regression: a practical and nonmathematical review. FASEB J 1987;1:365–74.
11. Baud M. Data analysis, mathematical modeling. In: Masseyeff RF, Albert W, Staines NA, eds. Methods of Immunological Analysis Volume 1: Fundamentals. New York: VCH Publishers Inc, 1993:656–71
12. Staros JV, Wright RW, Swingle DM. Enhancement by N-hydroxysulfosuccinimide of water-soluble carbodiimide-mediated coupling reactions. Anal Biochem 1986;156:220–2.
13. Grabarek Z, Gergely J. Zero-length crosslinking procedure with the use of active esters. Anal Biochem 1990;185:131–5.
14. Timkovich R. Detection of the stable addition of carbodiimide to proteins. Anal Biochem 1977;79:135–43.
15. de Jager W, Rijkers GT. Solid-phase and bead-based cytokine immunoassay: a comparison. Methods 2006;38:294–303.

Chapter 10

Purification of the T Cell Antigen Receptor and Analysis by Blue-Native PAGE

Mahima Swamy and Wolfgang W.A. Schamel

Abstract

The T cell antigen receptor (TCR) is a multi-protein complex composed of six different transmembrane subunits, which form complexes of various sizes on the surface of resting T cells. The stoichiometry of the smallest form was recently determined to be $\alpha\beta\gamma\varepsilon\delta\varepsilon\zeta\zeta$, whereas that of the larger forms is unknown. The roles of the different forms and their ratios are poorly defined. Biochemical analyses to address these questions must focus on the detergent and the best native conditions to maintain the integrity of the complexes. Blue-native polyacrylamide gel electrophoresis (BN-PAGE) is a high-resolution native protein separation method that relies on the dye Coomassie blue to confer negative charge for separation. Using this powerful approach, the size, subunit composition and the relative abundance of the different TCR forms can be studied. We present here four methods to isolate the TCR in a native form and details to analyse it by BN-PAGE.

Key words: T cells, T cell receptor, BN-PAGE, multi-protein complex, stoichiometry, affinity purification.

1. Introduction

Mature T cells specifically recognize antigenic peptide-major histocompatibility complex (pMHC) complexes through their T cell antigen receptors (TCR). This recognition event leads to activation of the T cells, and the triggering of an immune response. Since soluble pMHC complexes must be at least dimeric in order to activate the TCR *(1, 2)*, it has been widely proposed that two TCRs must be brought together by multimeric pMHC, in order for the TCR to get activated (clustering model). This model requires that the TCR be monomeric before activation. The conformational change model, where pMHC induces a conformational

Gennaro De Libero (ed.), *T Cell Protocols: Second Edition, vol. 514*

DOI 10.1007/978-1-60327-527-9_10 Springerprotocols.com

change in the cytoplasmic tails of the TCR *(3, 4)*, or the segregation model *(5)*, where activated TCRs are segregated from phosphatases and inactivating influences, are indifferent to the multimerization status of the TCR. Hence, a method that can clearly define the stoichiometry and structure of the TCR on the cell surface would be very beneficial for further studies on the receptor.

Biochemical studies that analysed the structure of the TCR on the cell membrane have often reached conflicting conclusions due to the complex nature of this transmembrane receptor (for example, see *6, 7*). The TCR is composed of eight transmembrane chains: the antigen-binding variable TCR$\alpha\beta$ heterodimer and the invariant CD3$\varepsilon\gamma$, CD3$\varepsilon\delta$, and TCRζ_2 (CD247$_2$) dimers (**Fig. 10.1a**). These dimers are non-covalently associated with

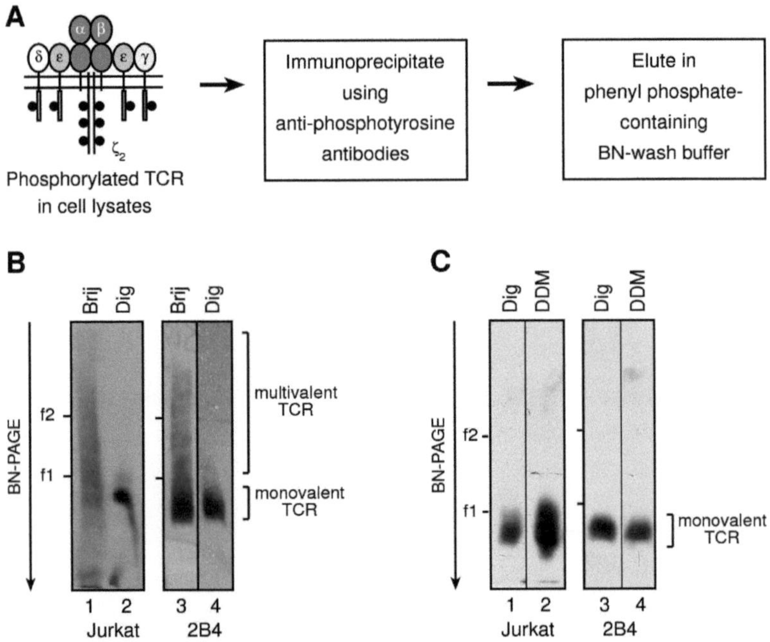

Fig. 10.1. Isolation of the TCR using anti-phosphotyrosine antibodies and analysis by BN-PAGE.
(**a**) Scheme for preparation of the TCR using anti-phosphotyrosine antibodies for immunoprecipitation and phenyl phosphate for elution. *Black circles* on the TCR indicate the sites of ITAM phosphorylation. (**b**) and (**c**) Pervanadate-treated cells, either Jurkat (human) or 2B4 (mouse) T cell lines, were lysed in either 0.5% Brij96, 1% digitonin or 1% DDM. After anti-phosphotyrosine immunoprecipitation and elution with 10 mM phenyl phosphate, the eluates were separated on BN gels. After the gel run, the proteins were transferred to PVDF membranes, and immunoblotted with an anti-TCRζ serum. (**b**) The TCR was detected in its monovalent (smaller) and multivalent (larger) complexes when purified using the detergent Brij (lanes 1 and 3), whereas all TCR is broken down to the monovalent form by the detergent digitonin (lanes 2 and 4). (**c**) The detergent DDM also breaks down all TCRs to the monovalent form (lanes 2 and 4). The bands of the marker protein ferritin in its 24-meric (f1) and 48-meric (f2) forms are indicated on the *left* of each panel.

each other in the context of the membrane. The invariant subunits each have an intracellular signalling-competent tail, with which they transmit the information of ligand-binding to the cytosolic signalling proteins. Receptor activation through the TCR$\alpha\beta$ subunits needs to be transmitted to the CD3 and TCRζ chains in spite of the weak associations. Thus, biochemical analysis of the TCR must be focused on maintaining these interactions, and maintaining the receptor in its most native conformation.

A high-resolution method for the characterization of native protein complexes is blue-native polyacrylamide gel electrophoresis (BN-PAGE) *(8–11)*. This technique makes use of the fact that the dye Coomassie Blue binds non-specifically to all proteins does not denature the proteins and is negatively charged. When this dye is added to the sample before electrophoresis, the protein complexes are separated based on the size of the complex (plus detergent/lipid micelle), and the negative charge of the bound Coomassie Blue.

When applied to the TCR, BN-PAGE analysis showed that the TCR was found in variously sized complexes, depending on the detergent used (**Figs. 10.1–10.4**) *(12, 13)*. On further experimentation, we realized that the TCR exists in a minimal $\alpha\beta\gamma\varepsilon\delta\varepsilon\zeta\zeta$ complex, and also in larger complexes, which were either maintained, or broken down to the minimal complex, depending on which detergent was used. Large TCR complexes

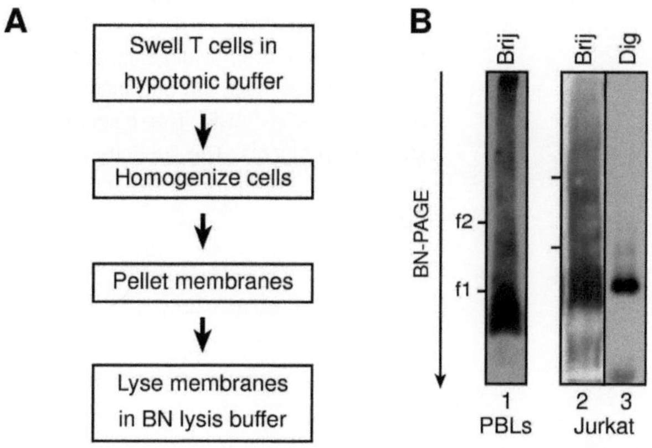

Fig. 10.2. Preparation of total cellular membranes and BN-PAGE analysis of the TCR. (**a**) Flow chart of the membrane preparation procedure. (**b**) Immunoblotting of the TCR in membrane preparations separated by BN-PAGE. Membranes were prepared by homogenizing unstimulated human peripheral blood lymphocytes. After ultracentrifugation, the membrane pellet was lysed in 0.5% Brij96, and analysed by BN-PAGE (lane 1). As in **Fig. 10.1**, multivalent and monovalent complexes were detected. Membrane preparations of the human T cell line Jurkat were lysed in either 0.5% Brij96 (lane 2) or 1% digitonin (lane 3), separated by BN-PAGE, and then immunoblotted with an anti-TCRζ serum. Again, digitonin disrupted the multivalent TCRs. The ferritin marker bands are indicated to the *left* (f1 and f2).

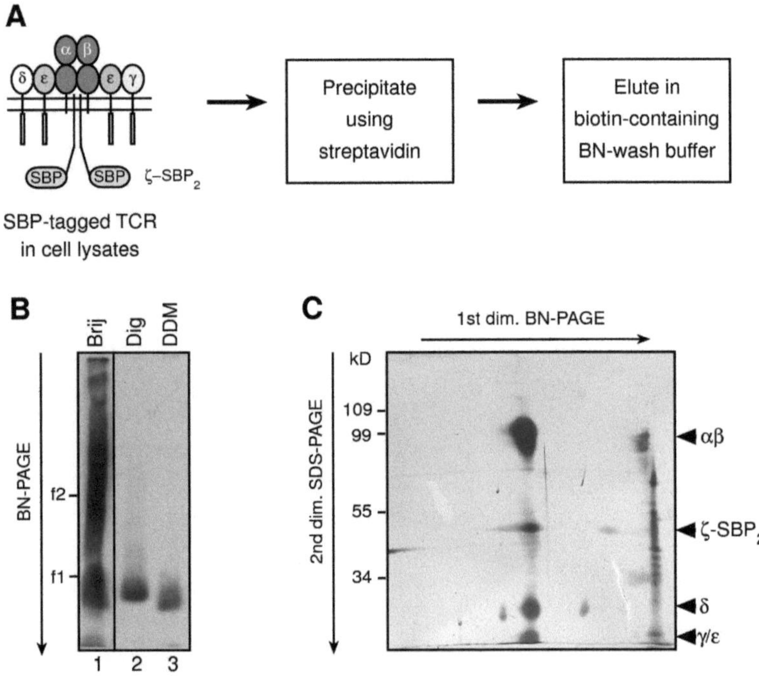

Fig. 10.3. One-step purification of the SBP-tagged TCR and analysis by BN-PAGE. (**a**) ζ-SBP containing TCR and the scheme for its purification procedure. (**b**) BN-PAGE analysis of the ζ-SBP TCR. Mζ-SBP murine T cells were lysed in either 0.5% Brij96 (lane 1), 1% digitonin (lane 2), or 1% DDM (lane 3) and the ζ-SBP TCR was precipitated using streptavidin–sepharose. After elution using 2 mM biotin, the TCRs were resolved by BN-PAGE. Immunoblotting was performed with an anti-ζ serum. (**c**) High yield and purity are obtained using the ζ-SBP TCR. The digitonin-extracted TCR from Mζ-SBP cells was purified using streptavidin–sepharose and biotin elution, and then separated by BN-PAGE. The lane was cut out, incubated in Laemmli buffer and subjected to a second dimension SDS-PAGE. The second dimension gel was then silver-stained as per standard protocols. The subunits corresponding to the spots on the gel are indicated to the *right*. The sizes of the marker protein bands are indicated to the *left*, in kilo Daltons.

were also detected by electron microscopy of immunogold-labelled TCRs on fixed T cell membranes. Thus, the TCR exists in larger clusters with more than one TCR$\alpha\beta$ before ligand-binding. Additionally, the ratio between large and small complexes varies from cell to cell. Several studies did not detect the TCR multimers, since they used detergents that disrupt these complexes; reviewed in **Ref.** *(13)*. Thus, we would like to stress that the choice of detergent for TCR analysis must be carefully thought over.

The existence of large TCR complexes containing up to ten TCR$\alpha\beta$ indicates that the clustering model of TCR activation becomes rather unlikely. New models must then be developed for the mechanism of T cell activation. We have recently suggested that TCR triggering requires two events simultaneously:

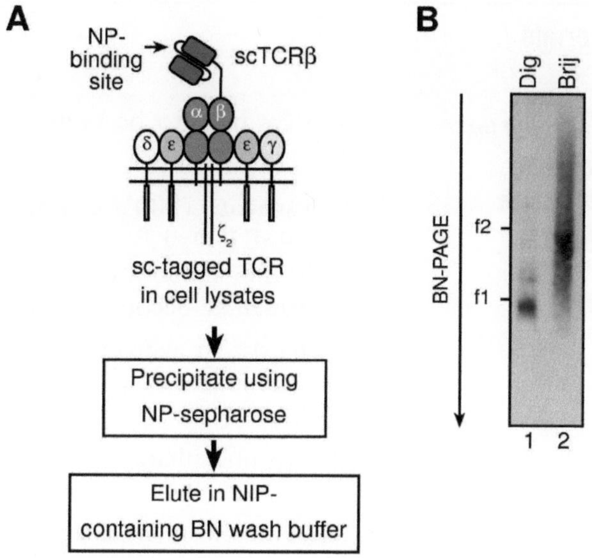

Fig. 10.4. One step purification of the sc-tagged TCR and analysis by BN-PAGE. (**a**) A schematic representation of the scTCR and the procedure for its purification are shown. (**b**) BN-PAGE analysis of the digitonin- (lane 1) and Brij96- (lane 2) extracted scTCR. 31.13scTCRβ cells were lysed in 1% digitonin or 0.5% Brij96 and the scTCR was affinity purified by NP-sepharose beads. The TCR was then eluted using 5 mM free NIP-CAP. After separation by BN-PAGE, immunoblotting was performed with an anti-TCRζ serum, as in previous figures.

TCR clustering (either induced by multimeric ligand-binding or present before ligand-binding) and a conformational change at CD3 (induced by multimeric ligand-binding) *(4)*. In addition, the existence of multivalent TCR clusters also has an impact on the sensitivity of T cells to antigen, as these clusters would have an increased avidity for multimeric pMHC complexes *(13)*.

We describe here the various isolation techniques that are used to obtain the TCR complex in a native state and the method for pouring and running BN-PAGE. The first technique involves immunoprecipitating the phosphorylated TCR and then eluting it from anti-phosphotyrosine beads using phenyl phosphate. This is our most preferred method of purification, due to ease of application, and the possibility to apply it to all cell types. For this method the TCR has to be phosphorylated. The second method, which is also universally applicable, is to prepare total cellular membranes. The last two methods describe the purification of the TCR through affinity-based tags. These methods have the caveat that they require cell lines expressing the tagged subunit of the TCR, but they do provide efficient and pure preparations of the TCR.

2. Materials

2.1. Isolation of the TCR Using Anti-Phosphotyrosine Antibodies

All buffers have to be prepared with distilled/deionized water (dH_2O).

1. Pervanadate (100 ×): 5 µl 50 mM sodium orthovanadate, 5.7 µl dH_2O, 1.5 µl 30% H_2O_2. Mix in this order; incubate 5–30 min at room temperature before use. Due to oxidation the vanadate becomes brownish.
2. Sepharose or agarose beads coupled to anti-phosphotyrosine antibodies (for example, the antibody 4G10, Upstate Biotech, UK).
3. Phenyl phosphate (prepare a 1 M stock in dH_2O just before use).
4. Alkaline phosphatase (e.g. from New England Biolabs).
5. Detergents:
   a. Digitonin: 2% w/v stock solution in dH_2O. Freeze aliquots at −20°C. Thawed solutions are stable at room temperature for up to a week. If a precipitate forms, reheat the solution to 95°C until the digitonin redissolves.
   b. Brij 96: 10% w/v stock solution in dH_2O. Store at room temperature.
   c. Dodecylmaltoside (DDM): 10% w/v stock solution in dH_2O. Store aliquots at 4°C.
6. Protease and phosphatase inhibitor stock solutions (toxic):
   a. Leupeptin (1,000 ×): Prepare a 10 mg/ml solution in dH_2O. Store 1 ml aliquots at 20°C.
   b. Aprotinin (1,000 ×): Prepare a 10 mg/ml solution in dH_2O. Store 1 ml aliquots at 20°C.
   c. PMSF (100 ×): Prepare a 100 mM solution in ethanol. Store 1 ml aliquots at 20°C.
   d. Sodium orthovanadate (100 ×): Prepare a 50 mM solution in dH_2O. Store at room temperature.
   e. Sodium fluoride (100 ×): Prepare a 1 M solution in dH_2O. Store at room temperature.
7. Lysis buffer (1 ×): 20 mM Tris–HCl, pH 7.4, 137 mM NaCl, 2 mM EDTA pH 8.0, 10% glycerol. Prepare a 2 × stock and store at 4°C. Add detergent of interest (end concentration: 1% digitonin or 0.5% Brij 96 or 1% DDM) and protease and phosphatase inhibitors just before use, at 1 × concentration.

8. BN lysis buffer (1 ×): 20 mM Bis–tris pH 7.0, adjusted with HCl, 500 mM 6-aminohexanoic acid, 20 mM NaCl, 2 mM EDTA pH 8.0, 10% glycerol. Prepare a 2 × stock and store at 4°C. Add detergent of interest and protease and phosphatase inhibitors as for lysis buffer. Upon addition of sodium orthovanadate the buffer will get a yellowish colour.
9. BN wash buffer (1 ×): Same as BN lysis buffer, but the amount of detergent is reduced to half. Add protease inhibitors and omit phosphatase inhibitors.

2.2. Preparation of Total Membrane Fractions

1. Hypotonic lysis buffer: 10 mM HEPES, pH 7.4, 42 mM KCl, 5 mM $MgCl_2$. Store at 4°C. Add protease inhibitors just before use.
2. Dounce homogenizer.
3. Ultracentrifuge (e.g. from Beckman TLA100).

2.3. Purification of the TCR Through Tagged Subunits of the Receptor

1. The streptavidin-binding-peptide (SBP) is a 38-amino acid sequence that can be utilized to purify attached proteins due to its high affinity (K_D=2.5 nM) to streptavidin-coupled beads *(14)*. The plasmid encoding TCRζ-SBP was generated by ligating the sequence of SBP to the 3'end of the coding sequence of the murine TCRζ chain through a flexible linker. Upon transfection of this DNA into the murine TCRζ-deficient cell line MA5.8, a TCR-SBP (**Fig. 10.3a**) expressing clone was obtained (Mζ-SBP).
2. Streptavidin-coupled agarose beads (for example, UltraLink Immobilized Streptavidin Plus Gel, Pierce Biotechnology, USA).
3. Biotin, 40 mM stock solution in 0.02 N NaOH. Store at 4°C.
4. The expression vector for the single chain-tagged TCRβ (scTCRβ) was generated by fusing the sequence for the single chain Fv fragment of an anti-NIP antibody to the sequence of the human Vβ HA1.7 chain. The cell line 31–13.scTCRβ was generated by transfecting an expression vector encoding for the scTCRβ protein into the TCRβ-deficient mutant 31–13 *(4)*, which is derived from the human T cell line Jurkat. The sc-tag was introduced C-terminal of the signal sequence cleavage site.
5. NP (4-hydroxy-3-nitrophenylacetyl) sepharose beads (Biosearch Technologies).
6. NIP-CAP (4-hydroxy-3-iodo-5-nitrophenylacetic caproic acid, Biosearch Technologies), 500 mM stock solution in DMSO. Store at room temperature.

2.4. Pouring and Running of the BN-PAGE

1. Acrylamide:bisacrylamide solution in the ratio 32:1, %T=41.25, %C=3 (*see* **Note 1**). Acrylamide is a potent neurotoxin, so suitable precautions must be taken while handling.
2. Bis–tris (e.g. from Roth).
3. Tricine.
4. 6-Aminohexanoic acid (also known as ϵ-aminocaproic acid).
5. Coomassie Blue G250 (Serva).
6. Gradient mixer (e.g. from Bio-Rad).
7. Silicon tubing, 3–5 mm diameter, 1 m length.
8. Peristaltic pump (e.g. from Amersham Pharmacia Biotech).
9. Gel electrophoresis system (e.g. Bio-Rad Protean II or III, *see* **Note 2**).
10. Semi-dry transfer equipment.
11. Polyvinylidene difluoride (PVDF) membrane (e.g. from Immobilon P, Millipore).
12. SDS-PAGE gel running system and buffers.
13. Silver staining and/or Coomassie staining reagents.
14. Anti-TCR antibody for immunoblotting (*see* **Note 3**).
15. BN-gel buffer (3 ×): 150 mM Bis–tris pH 7.0, adjusted with HCl, 200 mM 6-aminohexanoic acid. Store at 4°C.
16. Low percentage separating gel, 4% (adjust volumes as per requirements): 1.5 ml acrylamide:bisacrylamide solution, 5 ml BN-gel buffer (33 ×), 8.5 ml dH_2O, 54 μl APS (ammonium persulfate, 10% w/v in dH_2O), 5.4 μl TEMED.
17. High-percentage separating gel, 9% (adjust volumes as per requirements): 3.38 ml acrylamide:bisacrylamide solution, 5 ml BN-gel buffer (3 ×), 6.63 ml glycerol 70%, 42 μl APS (ammonium persulfate), 42 μl TEMED.

    Note that the combined volumes of the two solutions (low- and high- percentage separating gels) should be exactly equal to the volume required to fill the space between the glass plates to the required height.
18. Stacking gel, 3.2% (adjust volumes as per requirements): 0.72 ml acrylamide:bisacrylamide solution, 3 ml BN-gel buffer (3 ×), 5.28 ml dH_2O, 120 μl APS (ammonium persulfate), 12 μl TEMED.
19. Anode buffer: 50 mM Bis–tris pH 7.0, adjusted with HCl. Prepare a 10 × stock and store at room temperature. Buffer (1 ×) should be pre-chilled for use.
20. Cathode buffer: 15 mM Bis–tris pH 7.0, adjusted with HCl, 50 mM Tricine, 0.02% Coomassie Blue G250. Prepare a

10 × stock and store at room temperature. Buffer (1 ×) should be pre-chilled for use.

21. Marker protein mix: 10 mg/ml aldolase (158 kD), 10 mg/ml catalase (232 kD), 10 mg/ml ferritin (440 and 880 kD), 10 mg/ml thyroglobulin (670 kD), 10 mg/ml BSA (66 and 132 kD). Prepare in BN lysis buffer without detergent or inhibitors (*see* **Note 4**). Store at 4°C.

3. Methods

3.1. Isolation of the TCR Using Anti-Phosphotyrosine Antibodies

Pervanadate, generated by mixing vanadate and H_2O_2, is a tyrosine phosphatase inhibitor, and can thus be employed to induce maximal phosphorylation of the TCR. This phosphorylation by pervanadate does not change the stoichiometry of the receptor, nor does it affect the multimerization status. On the other hand, it provides a convenient method to isolate the TCR, as the TCR has at least 20 tyrosine phosphorylation sites in the immunoreceptor tyrosine-based activation motifs (ITAMs) (**Fig. 10.1a**), and can be immunoprecipitated with anti-phosphotyrosine antibodies. Subsequently, the TCR can be eluted from the beads using phenyl phosphate and then dephosphorylated using alkaline phosphatase. While this is the most convenient technique for the enrichment of the native TCR, it has the disadvantage that other phosphoproteins are also purified. This procedure can also be used for T cells stimulated by other methods that also induce phosphorylation of the TCR, for example, stimulation by pMHC.

1. Harvest 30 × 10^6 T cells and resuspend in 1 ml serum-free medium.
2. Equilibrate to 5% CO_2, 37°C, for 10 min, and then stimulate with 10 μl pervanadate for 5 min.
3. Pellet the cells briefly (1,000 *g*), and resuspend in 1 ml lysis buffer containing the appropriate detergent at 4°C.
4. Lyse the cells for 15 min on ice, and then centrifuge the lysate at 12,000 *g* for 15 min at 4°C.
5. Add 3–10 μl of anti-phosphotyrosine beads to the clarified supernatant.
6. Incubate the clarified supernatant with the beads for at least 4 h by rotating at 4°C.
7. Pellet the beads briefly (1,000 *g*), and wash the beads three times with BN wash buffer containing the same detergent as the lysis buffer.

8. Prepare the elution buffer from the BN wash buffer, by adding 100 mM phenyl phosphate to it.
9. Elute the TCR (along with other phosphorylated proteins) from the beads at 4°C, by resuspending the beads in 20–100 μl of BN elution buffer.
10. Incubate the beads in the elution buffer for 30 min at 4°C, resuspending occasionally (*see* **Note 6**).
11. To dephosphorylate the TCR, 1–2 units of alkaline phosphatase can be added during the last 5 min of the incubation.
12. Pellet the beads briefly, and remove the supernatant to a new tube.
13. The TCR-containing eluate can now be separated by BN-PAGE.

3.2. Preparation of Total Membranes for Analysis of the TCR

In certain experiments pervanadate stimulation of the TCR is not desirable. In these cases, total membrane preparations can be made for analysis by BN-PAGE, which has the additional advantage that the yield is high. Briefly, the cells are allowed to swell in a hypotonic buffer, and then broken by mechanical shearing. The resultant membrane patches are pelleted in an ultracentrifuge, and then lysed in the BN buffer (**Fig. 10.2**). Since this method does not separate the plasma membrane from other membranes of the cell, internalized TCRs, and partially assembled TCRs are also isolated. Also, as with purifications using anti-phosphotyrosine antibodies, other proteins are purified along with the TCR.

1. Wash 5–15 $\times 10^6$ T cells in cold PBS, and resuspend in 1 ml hypotonic lysis buffer.
2. Allow the cells to swell for 10 min on ice and then break the cells by mechanical shearing in a Dounce homogenizer, 10–15 strokes, for a total of 3 min.
3. Centrifuge the homogenate for 10 min at 150 *g* at 4°C to get rid of whole cells and nuclei.
4. Pellet the membrane patches containing the TCR by centrifuging the supernatant at 100,000 *g* for 30 min in an ultracentrifuge at 4°C.
5. If cytoplasmic contaminants are to be stringently removed, wash the membrane pellet once in hypotonic lysis buffer, and centrifuge at 100,000 *g* for 10 min.
6. Resuspend the membrane pellet carefully, in 250 μl BN lysis buffer containing the appropriate detergent at 4°C. Incubate for 45 min to 1 h, periodically resuspending with a pipette, to ensure that the membranes are completely dissolved.
7. Centrifuge this membrane lysate in an ultracentrifuge for 10 min at 100,000 *g* at 4°C (*see* **Note 5**).

8. The supernatant is the TCR-containing membrane lysate that can be separated by BN-PAGE.

3.3. Purification of the TCR Through Tagged Subunits of the Receptor

If a very pure preparation of the TCR is needed for analysis by BN-PAGE, the best method is to purify a tagged version of the receptor. We have used two different tags; the streptavidin-binding peptide (SBP) *(14)* fused to the C-terminus of the TCRζ chain (ζ-SBP, **Fig. 10.3a**), and a single chain variable fragment Fv (sc) of an anti-NIP antibody *(15)* attached to the N-terminus of TCRβ (scTCRβ, **Fig. 10.4a**). The SBP-tag can bind to streptavidin, and can thus be bound to streptavidin supports. As biotin has a higher affinity for streptavidin, it releases the SBP-tag from the streptavidin. Similarly, the sc tag has a higher affinity for nitro-iodo-phenol (NIP) than for nitro-phenol (NP), and can thus be bound to NP-sepharose beads, and eluted with soluble NIP. Both of these affinity-tag based purification methods require expression of the tagged subunit, preferably in a cell line that lacks the corresponding endogenous subunit. While this requirement is indeed a drawback, these tags are useful for highly efficient purification of the TCR.

3.3.1. Purification of the TCR Utilizing ζ-SBP

1. Harvest 20 $\times$ 10^6 cells expressing TCRζ-SBP (Mζ-SBP) and wash once in PBS.
2. For purification of ζ-SBP-containing protein complexes, lyse the cells in 1 ml ice-cold lysis buffer as in **Section 3.1**, steps 3–4.
3. Add streptavidin-coupled agarose or sepharose (approximately 7 µl) to the clarified supernatant, and allow binding of the TCR-SBP for 2–4 h rotating at 4°C.
4. Wash beads three times in BN-wash buffer.
5. Elute the TCR-SBP from the beads by incubation with 100 µl BN wash buffer containing 2 mM biotin for 20 min at 4°C. Resuspend occasionally with a pipette (*see* **Note 6**).
6. Centrifuge the beads briefly, and transfer the supernatant to a new tube.
7. The TCR-containing eluate is now ready to be separated by BN-PAGE.

3.3.2. Purification of the TCR Utilizing scTCRβ

1. Harvest 20 $\times$ 10^6 of the 31.13scTCRβ cells. Lyse the cells as given before (**Section 3.1**).
2. To purify the scTCR through the sc fragment of the anti-NIP antibody, allow the cleared lysate to bind to 5 µl NP-sepharose beads for 4 h.
3. Wash the beads as above, and then elute the TCR from the beads in 20 µl BN wash buffer containing 5 mM NIP-CAP.

Resuspend the beads occasionally, and elute for 30 min (*see* **Note 6**).

4. Pellet the beads, and transfer the eluate to a new chilled tube.
5. The TCR-containing eluate is now ready to be separated by BN-PAGE.

3.4. Pouring and Running of the Blue-Native PAGE Gels

BN-PAGE gels are polyacrylamide gradient gels to ensure better separation over a wide range of protein complexes size *(9)*. The high-percentage separating gel solution contains glycerol, so that it settles down due to its higher density and aids in the gradient formation. The gels are poured at room temperature, and pre-chilled before loading. The running of the gel is then completely performed at 4°C. A protocol to pour and run small (6 × 10 cm, 1 mm thick, e.g. Protean II Bio-Rad) gels is given *(11)*.

1. The set-up for the pouring of the gels is shown in **Fig. 10.5**. Place the gradient mixer on the magnetic stirrer, and the magnet in the chamber closer to the exit. Attach the tubing as shown (**Fig. 10.5**), and pass it through the peristaltic pump. At the other end, attach a syringe needle, and place the tip of the needle between the glass plates of the gel apparatus. Keep the connecting channel between the two

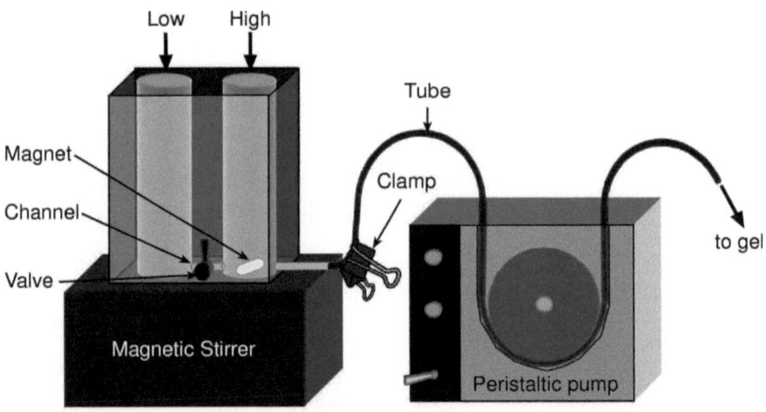

Fig. 10.5. Schematic representation of the set-up for the pouring of the native gradient gels.
The gradient mixer is placed on the magnetic stirrer with a magnet stir bar in the more proximal chamber. The tube from the gradient mixer is passed through a peristaltic pump, which controls the flow of the gel liquid into the gel apparatus. The high- and low-percentage gel mixtures should be poured into the indicated chambers of the gradient mixer. Once the gel mixtures are in the chambers, the stirrer is switched on, the channel between the chambers is opened by the valve, the clip is removed and the pump is switched on. The gel mixture is allowed to slowly enter between the glass plates of the gel apparatus (not shown). When all the liquid has entered the gel apparatus, the gel is overlaid with isopropanol and allowed to polymerize.

chambers of the mixer closed (using the valve) as well as the out-flowing tubing (using the clamp).

2. Prepare the low- and high-percentage separating gel solutions, adding APS and TEMED immediately before use.
3. Pour the low- and high-percentage solutions into the corresponding chambers of the mixer. Switch on the magnetic stirrer and open the channel valve. Remove the clamp and switch on the peristaltic pump at a speed that allows the gel solution to fill in between the glass plates in 1–2 min. Ensure that liquid is flowing between the two chambers of the gradient mixture by putting gentle pressure on the side of the "low" chamber.
4. Allow all the solution to enter between the glass plates, always maintaining the needle tip above the liquid level. Overlay with isopropanol. Allow the gel to polymerize for at least 30 min at room temperature.
5. Remove the isopropanol, rinse with water, air-dry (*see* **Note** 7).
6. Pour the stacking gel and insert the combs as for a normal SDS Laemmli gel.
7. Allow the gel to polymerize for 10 min at room temperature, and for a further 20 min at 4°C. This helps to make the stacking gel slightly more solid, which aids in the removal of the comb.
8. Remove the comb slowly, moving it back and forth against the plates, so that air can enter into the wells. The upper gel is semi-solid, so this improves the quality of the wells. Maintain the gel at 4°C.
9. Rinse the wells with dH_2O, and then load the prepared samples in the dry wells in a cold room. Load 10 μl of the marker protein mix.
10. Overlay the sample in each well with cold cathode buffer.
11. Fit the gel in the running apparatus; fill the inner chamber with cold cathode buffer, and the outer/lower chamber with cold anode buffer.
12. Allow the samples to enter the stacking gel at 20–50 V, after which the gel can either be run overnight at 20 V, or at 2 mA/ cm^2 until the running front has reached the bottom of the gel. In the second case, the gel run takes between 3 and 4 h. The entire procedure is done in a cold room/ at 4°C. After the gel run, the gel will be completely blue, with a more intense blue running front.
13. Transfer the proteins from the BN-PAGE gel to a PVDF membrane using standard blotting procedures. The transfer

buffer must contain 0.1% SDS for efficient subsequent immunoblotting (*see* **Note 8**).

14. Silver or Coomassie staining are also possible.
15. The individual lanes of the BN-PAGE gel can also be cut out and loaded on a two-dimensional SDS-PAGE. Incubating the gel strip in Laemmli buffer beforehand for 30 min ensures complete denaturing of the proteins (*see* **Notes 9 and 10**). After two-dimensional BN-/SDS-PAGE, normal Western blotting, silver or Coomassie staining can be done (*see* **Fig. 10.3c**).

4. Notes

1. We prepare acrylamide:bisacrylamide solution with the right ratio of 32:1 by mixing 17.66 ml Rotiphorese Gel 40 (40% acrylamide, Roth, Germany) that has an acrylamide:bisacrylamide ratio of 19:1 with 82.34 ml of Rotiphorese Gel 40 with a ratio of 37.5:1.
2. Maintain all apparatus for the BN-PAGE SDS-free. Rinse only in dH_2O to clean.
3. Not all antibodies work for immunoblotting after BN-PAGE. This could be due to the bound Coomassie, which can only be partially removed by methanol from PVDF membranes, and not at all from nitrocellulose. Additionally, fluorescence detection of proteins on the membrane is not possible, because Coomassie is highly fluorescent (in both red and green channels).
4. Of the marker proteins listed here, only ferritin is visible during the gel run due to its brown colour. Since the monovalent TCR is normally found just below the height of the first ferritin band, we routinely use only this protein as a marker in our gels.
5. If total membranes are prepared, there is always some amount of precipitation of Coomassie and proteins in the well, and also at the interface between the stacking and separating gel. If, however, there is excessive precipitation, so that proteins are not able to enter the well (seen as streaks in the gel where the blue dye does not enter), then a second ultra centrifugation step after lysis of the membrane pellets may be necessary.
6. To increase TCR recovery during elution from the affinity columns/beads from small cell numbers (**Sections 3.1 and 3.3**), two rounds of elution can be performed serially using the same beads. Subsequently, the eluates are pooled.

7. After the lower separating part of the BN-PAGE gel has been poured, it can be overlaid with 1 × BN-gel buffer, and can then be stored for up to half a year in a humid environment at 4°C. We routinely pour 12–13 gels using the Bio-Rad multi-caster system, and store the gels for use as required.
8. Standard procedures are used for transferring the proteins from the native gels to PVDF or nitrocellulose membranes. Inclusion of SDS in the transfer buffer can help for immunoblotting of the TCR with antibodies that recognize the denatured proteins. Conversely, native transfer conditions (without SDS) may aid detection by antibodies that recognize the native proteins (*see also Note 3*).
9. If performing a second dimension SDS-PAGE, it helps to put tape (i.e. Cello tape or Tesafilm) on the spacers of the SDS-PAGE gel apparatus to slightly increase the thickness of the gel. Then the BN gel slice can easily slide in between the two glass plates, but is still held relatively tightly in place.
10. If performing a second dimension SDS-PAGE, boiling of the gel slice in Laemmli buffer is normally not necessary for the TCR; however, if recovery in the second dimension gel is not very good, briefly boiling the gel slice (20 s) in SDS-containing buffer may help.

Acknowledgements

We thank Michael Reth for his scientific support and acknowledge the contributions of Gabrielle Siegers, Susana Minguet, Elaine Dopfer, Sandra Ernst and Thomas Bock. This work was supported by an Emmy Noether Fellowship and grant SFB620 from the DFG to W.W.S.

References

1. Boniface JJ, Rabinowitz JD, Wulfing C, et al. Initiation of signal transduction through the T cell receptor requires the multivalent engagement of peptide/MHC ligands [corrected]. Immunity 1998;9:459–66.
2. Cochran JR, Cameron TO, Stern LJ. The relationship of MHC-peptide binding and T cell activation probed using chemically defined MHC class II oligomers. Immunity 2000;12:241–50.
3. Gil D, Schamel WW, Montoya M, Sanchez-Madrid F, Alarcon B. Recruitment of Nck by CD3 epsilon reveals a ligand-induced conformational change essential for T cell receptor signaling and synapse formation. Cell 2002;109:901–12.
4. Minguet S, Swamy M, Alarcon B, Luescher IF, Schamel WW. Full activation of the T cell receptor requires both clustering and conformational changes at CD3. Immunity 2007;26:43–54.
5. Choudhuri K, Kearney A, Bakker TR, van der Merwe PA. Immunology: how do T cells recognize antigen? Curr Biol 2005;15:R382–5.

6. Punt JA, Roberts JL, Kearse KP, Singer A. Stoichiometry of the T cell antigen receptor (TCR) complex: each TCR/CD3 complex contains one TCRa, one TCRb and two CD3e chains. J Exp Med 1994;180:587–93.
7. Fernandez-Miguel G, Alarcon B, Iglesias A, et al. Multivalent structure of an alphabetaT cell receptor. Proc Natl Acad Sci USA 1999;96:1547–52.
8. Schägger H, Cramer WA, von Jagow G. Analysis of molecular masses and oligomeric states of protein complexes by blue native electrophoresis and isolation of membrane protein complexes by two-dimensional native electrophoresis. Anal Biochem 1994; 217:220–30.
9. Schägger H, von Jagow G. Blue native electrophoresis for isolation of membrane protein complexes in enzymatically active form. Anal Biochem 1991;199:223–31.
10. Swamy M, Kulathu Y, Ernst S, Reth M, Schamel WW. Two dimensional Blue Native-/ SDS-PAGE analysis of SLP family adaptor protein complexes. Immunol Lett 2006; 104:131–7.
11. Swamy M, Siegers GM, Minguet S, Wollscheid B, Schamel WW. Blue native polyacrylamide gel electrophoresis (BN-PAGE) for the identification and analysis of multiprotein complexes. Sci STKE 2006;2006:pl4.
12. Schamel WW, Arechaga I, Risueno RM, et al. Coexistence of multivalent and monovalent TCRs explains high sensitivity and wide range of response. J Exp Med 2005;202: 493–503.
13. Alarcon B, Swamy M, van Santen HM, Schamel WW. T-cell antigen-receptor stoichiometry: pre-clustering for sensitivity. EMBO Rep 2006;7:490–5.
14. Keefe AD, Wilson DS, Seelig B, Szostak JW. One-step purification of recombinant proteins using a nanomolar-affinity streptavidin-binding peptide, the SBP-Tag. Protein Expr Purif 2001;23:440–6.
15. Schamel WW, Kuppig S, Becker B, Gimborn K, Hauri HP, Reth M. A high molecular weight complex of BAP29/BAP31 is involved in the retention of membrane-bound IgD in the endoplasmic reticulum. Proc Natl Acad Sci 2003;100:9861–6.

Chapter 11

Non-Replicating Recombinant Vaccinia Virus Expressing CD80 to Enhance T-Cell Stimulation

Paul Zajac

Abstract

The following method describes the generation of a recombinant vaccinia virus expressing a costimulatory molecule (human CD80 or B7.1).

The procedure first requires the cloning, by classical methods not described here, of the gene of interest, e.g. CD80, into a vaccinia shuttle plasmid under the control of a virus-specific promoter enabling a transcription during the early phase of infection. Flanking the insert, the plasmid contains viral sequences and a selection maker needed for the insertion into the viral genome. The successive plaque isolation of recombinant virus on cell monolayer described here is based on the transient "gpt" selection system which enables other insertions in different loci of the same virus. Finally, after verification amplification and titration of the recombinant vector, replication will be impaired by a psoralen-UV treatment in order to produce a non-replicating virus. Expression and function of inserts, following infection of cells, are verified by specific phenotypic and functional assays.

Key words: Recombinant vaccinia virus, vaccine vector, CTL stimulation, costimulation, non-replicating viral vector.

1. Introduction

The induction of specific cytotoxic T lymphocytes (CTL) in vivo requires highly immunogenic vaccine formulation. Therefore, vaccines require adjuvants able to trigger "danger signals" *(1)* or able to directly provide immuno-modulators necessary for T-cell stimulation.

Recombinants vaccinia virus (rVV) are thermostable and relatively easy to produce. In addition, they are endowed with a number of interesting features. First, they can accomodate large inserts, including different genes *(2, 3)*. Second, high levels of expression of

Gennaro De Libero (ed.), *T Cell Protocols: Second Edition, vol. 514*

DOI 10.1007/978-1-60327-527-9_11 Springerprotocols.com

recombinant proteins can be easily obtained *(4)*. Third, integration into the host's genome does not occur, resulting in high safety of this vector system *(5)*. Fourth, rVV was shown to efficiently deliver recombinant epitopes into the class I antigen presentation pathway *(6)*. Infection of experimental animals with rVV vectors elicited CTL populations recognizing both VV antigens and peptides encoded by the inserted genes *(7)*.

Prior to any further considerations, we should underline that replicative vaccinia virus handling requires a biosafety level 2 facility. Some countries recommend vaccination of the persons in direct contact with the virus. In order to decrease potential risk of replicating viral vectors in vivo, replication incompetent viruses were developed, including genetically impaired vectors and non-human poxviruses *(4)*. Treatment of VV with psoralen and ultraviolet-light irradiation was shown to induce cross-linking of viral DNA and abrogation of cytopathic effect and viral replication. Despite complete inactivation and absence of pathogenicity, high-gene expression following infection with rVV can still be obtained provided that the inserted genes are driven by early promoters *(8)*.

Generation of such viral vectors combining antigens and cofactors *(9, 10)* has lead to an increasing number of clinical trial (*see* review by Kaufman *(11)*) confirming poxviruses formulation as one of the most effective vaccine formulations for immunotherapeutic approaches currently available.

2. Materials

1. CV-1 cell line : Monkey African Green kidney fibroblast, ATCC CCL70, ECACC No 87032605
2. Vaccinia Virus (wild-type Copenhagen strain obtained from Dr. Drillien Strasbourg, France)
3. cDNA of gene of interest, e.g. human CD80 was obtained from Pr. Dellabona, Milan, Italy)
4. Vaccinia shuttle plasmid used here was originally developed and obtained from Dr. Kangla Tsung (San Francisco, CA) and targets insertions in vaccinia A44L locus.
5. LipofectamineTM (Invitrogen, Paisley, UK)
6. OPTI-MEM®I medium (Invitrogen, Carlsbad, CA).
7. DMEM (and 2 × DMEM) + 10% Foetal bovine serum + antibiotic (DMEM-FCS or 2 × DMEM)
8. Mycophenolic acid (400 ×): 10 mg/ml in 0.1 N NaOH

9. Xanthine (40 ×): 10 mg/ml in 0.1 N NaOH
10. Hypoxanthine (400 ×): 10 mg/ml in H_2O
11. Agarose
12. Tris–HCl, 10 and 1 mM, pH 9
13. Sucrose
14. QIAmp® DNA Blood Mini Kit (Qiagen, Hilden, Germany)
15. PCR reagent
16. Vaccinia specific oligonucleotides
17. Neutral red (10 mg/ml in water)
18. Cristal violet solution: 0.1% in 20% ethanol
19. Psoralen (4'-aminomethyl trioxsalen, 100 μg/ml – Calbiochem, LaJolla, CA)
20. Stratalinker + 365 nm bulbs
21. Cup sonicator.
22. Human CD80 antibody

3. Methods

3.1. Cloning of Gene of Interest in Vaccinia Shuttle Plasmid

The human CD80 cDNA was originally cloned from human PBMC mRNA in a pUC derived plasmid. Using compatible restriction site, the gene was sub-cloned into the plasmid containing sequences from the vaccinia A44L locus (**Fig. 11.1**). Between the two A44L regions, the cloning cassette contains viral early promoters followed by multiple cloning sites followed by a transcriptional termination sequence.

Of note this sequence "TTTTTNT" recognized as a stop signal sequence for vaccinia early transcription should not be present in the recombinant gene. In the case of the human CD80 gene, one stop signal was found at position 766 and was modified using the "QuickChange™" Site-Directed Mutagenesis Kit (Stratagene, Cambridge, UK).

3.2. Infection/Transfection (see Note 1)

1. In a 35 mm petri dish, add 5 × 10^5 CV-1 cells in 2 ml DMEM-FCS.
2. Once the cells are adherent, remove the medium and infect with the wild-type virus at a ratio of 1 pfu (plaque forming unit) for 10–20 cells (MOI of 0.1/0.05) in a total volume of 300 μl PBS.
3. Gently shake the petri dish in order to cover the monolayer, place it in a 37°C incubator for 1 h (repeat shaking every 15 min).

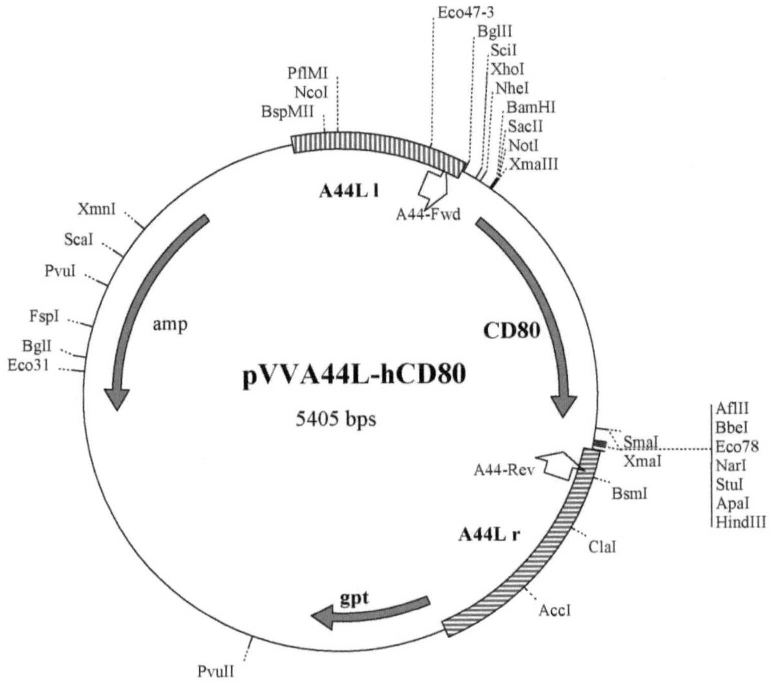

Fig. 11.1. Plasmid map of pVVA44L–hCD80.
Human CD80 cDNA was subcloned, between vaccinia early promoter and termination signal, in the EcoRI site of the cloning cassette which is flanked by sequences from vaccinia A44L locus (for homologous recombination). The primers A44-fwd and A44-rev enabling the amplification of the modified region can also be used for sequencing of the insert.

4. Forty-five minutes after the infection, prepare the transfection mix (avoid polypropylene plastic tube). In a first tube add 1–2 μg of plasmid to 100 μl OPTI-MEM. In a separate tube, add 10 μl lipofectamine to 100 μl of OPTI-MEM and mix gently with the plasmid suspension.
5. After 15 min incubation at room temperature, transfection mix is diluted in 1 ml OPTI-MEM and added onto the infected cells.
6. The transfected–infected cells are placed back at 37°C for 6–15 h, before adding 1 ml of 10% FCS-DMEM.
7. The infection is left in the incubator until complete lysis of the monolayer (2–3 days).
8. Transfer the 2 ml of viral solution and cell debris into a tube and keep at –20°C until the next step.

3.3. Selection of Recombinant Virus (see Notes 2–5)

Selection of the recombinant virus is based on *Escherichia coli* guanine phosphoribosyl transferase (gpt) *(12)* as a transient marker. In the shuttle plasmid, this "gpt" rescuing gene is located outside of the "stable recombinant zone" between the viral sequences leading to a transient unstable recombinant form. The final selection steps lead to the removal of the selection marker from the final rVV (**Fig. 11.2**).

1. Prepare one 6-well plate with 5×10^5 CV-1 cells per well (in 2 ml DMEM-FCS) and place at 37°C overnight in order to have a confluent monolayer at the time of infection.
2. Pre-incubate the culture at least 4 h with selective medium : DMEM-FCS + 1/400 MPA + 1/40 xanthine + 1/400 hypoxanthine

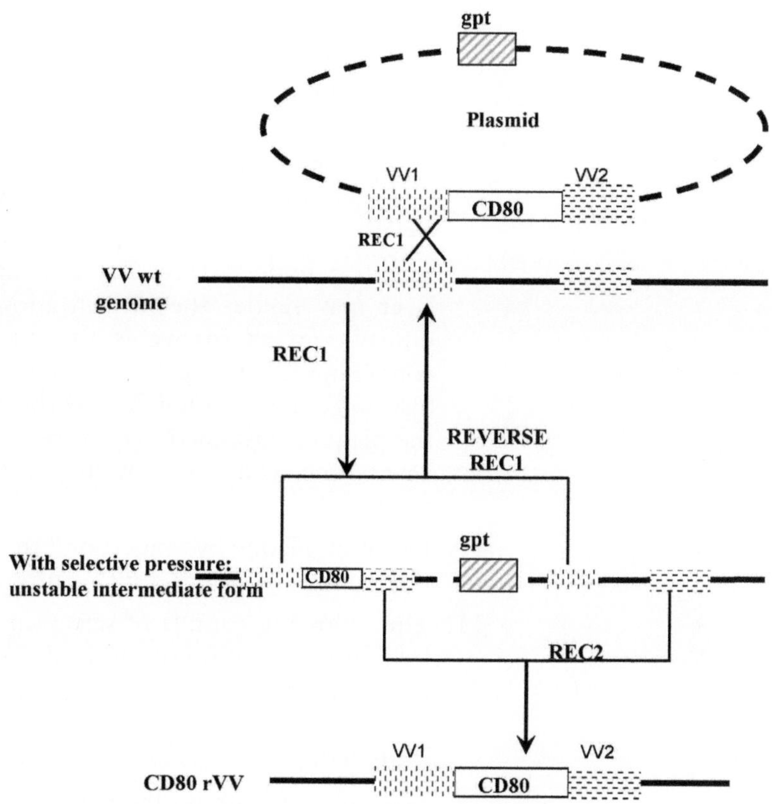

Fig. 11.2. Principle of transient selection marker for recombinant virus.
The cloning cassette is flanked by two sequences identical to viral regions (VV1 and VV2) allowing homologous recombinations by genetic crossing-over leading to the insertion of the gene of interest into the wild-type genome of Vaccinia virus.
During viral replication, a single homologous recombination event, e.g. with the region 1(REC1) generates a recombinant virus containing the entire plasmid whose presence in the viral genome is needed for selection under "pressure" by mycophenolic acid (MPA). However, this intermediate recombinant form contains duplicate sequences and is therefore genetically unstable. Removal of selective pressure will allow the isolation of stable viruses derived from this intermediate form following recombination events either with the two VV1 sequences reverting to the initial form (wt) or, following recombination with the second site (REC2) generating the stable recombinant virus (CD80rVV). The probability of each event is similar, thus the chances of obtaining the wt or the rVV are in theory 50/50.

3. Sonicate the closed tube containing the viral solution in the cup sonicator for 1 min in order to release the virus from intact cells and aggregates.
4. Prepare tenfold serial dilutions (200 μl in 1,800 μl PBS) until 10^{-4}
5. Remove the medium from the cells and infect two wells with 200 μl from the dilutions 10^{-2}, 10^{-3} and 10^{-4}. Place at 37°C for 1 h (shake gently every 15 min).
6. Prior to the end of the infection, melt 10 ml of 2% agarose solution and cool it to 45°C in a waterbath.
7. Prepare 10 ml of 2 × selective medium with 2 × DMEM-FCS + 50 μl MPA + 500 μl xanthine and 50 μl hypoxanthine. Mix with the 10 ml 45°C agarose just prior the next step
8. Remove inoculum from the cells and add 3 ml of agarose-selective medium. Leave at room temperature until solidification then transfer in a CO_2 incubator at 37°C.
9. After 2 days, mark viral plaques by circling (from under) with pen under a reverse microscope. Because the agarose layer may render the identification of viral plaques more difficult, an alternative is to add 2 ml of fresh agarose containing 1/100 of Neutral Red solution which will diffuse and stain (in a few hours) the cell monolayer and leave the plaques unstained. Try to select six plaques in order to have "one clone per well" in the following selection rounds.
10. Pick each plaque by "pipetting" the circled zone with sterile filtered tips and transfer into a tube containing 1 ml PBS.
11. The following rounds of selection will be performed similarly after sonication of the 1 ml PBS. It is recommended to perform at least two cycles of plaques isolation under selective medium followed by three cycles without pressure (*see* **Note 6**).
12. After the fifth cycle, perform a last infection with all the clones but under normal DMEM-FCS (no agarose) and leave the infection until complete lysis of the monolayer (2–3 days). The resulting 2 ml of viral solution, transferred in a tube, represents the so-called "mini-amplification" of each clone and will be used to identify the recombinant clones and as "seed" for the generation of larger stocks.

3.4. Identification of Recombinant Viral Clone

In order to verify the identity of the isolated viral clones, PCR is performed on the targeted viral sequences from purified vv genome. Primers specific for the flanking viral sequences are used in order to distinguish the unmodified wild-type vv genome from the insertion–deletion modified locus.

1. vv DNA purification from the clones "mini-amplifications" (as control for the PCR it is recommended to prepare DNA from cells infected with the wild-type virus) is performed by using DNA purification kit (QIAmp® DNA Blood Mini Kit, Qiagen), according to the manufacturer protocol. After thawing and sonication the 2 ml of each viral suspension, 200 μl are transferred in a 1.5 ml tube and mixed together with the lysis buffer and protease. The total DNA is eluted in 100 μl H_2O.
2. Total DNA purified from cells infected with each clone can now be used in a PCR reaction with specific primers for the vaccinia A44L locus :
   - Total DNA, 5 μl
   - Forward primer 5'-gaaagtaaactattatgg-3' (10 μM), 1 μl
   - Reverse primer 5'-gttgaaatctagttctgc-3' (10 μM), 1 μl
   - PCR buffer (10 ×), 2 μl
   - dNTP mix (10 mM), 2 μl
   - Taq DNA Pol, 1 μl
   - H_2O, 8 μl

PCR
Start : 95°C – 5 min
Cycle (× 30): 95°C – 15 s/56°C – 50 s/72°C – 60 s
End : 72°C – 5 min/4°C – standby

PCR amplicon are analysed on a 1% agarose gel (**Fig. 11.3**). Expected size of WT-A44L = 699 bp and of CD80-rVV-A44L =1,233 bp

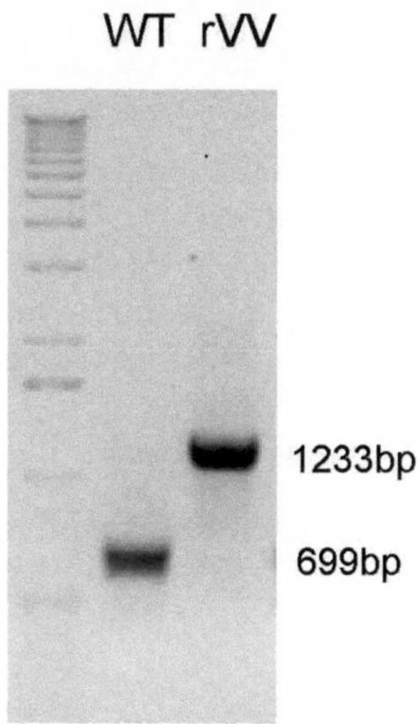

Fig. 11.3. Gel electrophoresis of PCR amplicons from A44L locus from VV genome.

Make sure that no wild-type band appears in the selected recombinant clone as this may be a sign of viral mixture requiring further plaque isolation cycles.

Moreover, to verify that the genome of the rVV clone does not contain anymore the resistance gene one can perform a comparative infection with and without selective medium. Plaques should only appear in infection without MPA.

3.5. Amplification and Concentration of rVV

The selected recombinant clone mini-amplification can now serve as "seed" to generate stocks of concentrated virus. The purification procedure is based on the increased density of concentrated sucrose solutions prepared in Tris 10 mM (weight/volume) and sterilized by filtration.

1. Prepare ten sub-confluent flasks (175 cm^2) of CV-1 cells
2. Infect with the selected rVV clone. Dilute 100 μl of the seed virus in 10 ml of PBS and directly add 1 ml in the culture medium of each flask.
3. Leave at 37°C until cytopathogenic effect (CPE) is complete (usually 3–4 days)
4. Detach the infected cells with one cycle of freeze–thaw and pool the cells by centrifugation (2,000 rpm 3 min) in a 50-ml tube. Pool and save the supernatants.
5. Resuspend the cell pellet in 10 ml Tris 10 mM and sonicate for 2 min. Centrifuge at 2,000 rpm for 3 min and transfer the supernatant in a fresh tube.
6. This step can be repeated a second time.
7. Prepare 15 ml of a 36% sucrose cushion in polyallomer ultracentrifuge tubes for SW-27 rotor (Beckman Coulter, Fullerton, CA), and overlay with 20 ml of viral supernatant (alternatively one can underlay the sucrose below the supernatant).
8. Centrifuge at 30,000 × *g* (13,000 rpm) for 90 min and resuspended the pellets in 3–5 ml Tris 1 mM.
9. This semi-purified viral solution can either be aliquoted and stored at –20°C or further purified on a sucrose gradient.
   a. In a SW-27 centrifuge tube, prepare a sucrose gradient by carefully overlaying 6 ml of sucrose 40, 36, 32, 28 and 24% before adding the virus and centrifuge at 26,000 *g* for 45 min.
   b. Collect the viral band in the middle of the continuous gradient established during centrifugation.
   c. Resuspend in 30–35 ml (final volume) with Tris 1 mM and pellet in a SW-27 sterile tube, by centrifugation at 30,000 *g* for 60 min.
   d. Collect the viral pellet with 2 ml Tris 1 mM and aliquot in freezing tubes.
   e. Store at –70°C (stocks) and keep one aliquot for the next steps.

3.6. Titration of rVV

Titer of virus is expressed as "plaque forming unit per millilitre of solution" = pfu/ml

1. Prepare one 6-well plate of confluent CV-1 cells.
2. Prepare a tenfold serial dilution in PBS of the sonicated stock virus until dilution 10^{-8} (100 μl in 900 μl PBS).
3. Remove medium from culture (leaving about 200 μl).
4. Add 100 μl of dilutions 10^{-6}, 10^{-7} and 10^{-8} in two wells.
5. Incubate at 37°C for 1 h with gentle agitation every 15 min.
6. Add 2 ml of medium and incubate 2 days at 37°C.
7. Replace the medium with 300 μl of crystal violet for 2 min.
8. Remove the staining solution and count the plaques.

   Calculate the concentration of the viral stock based on sufficient number of plaques (avoid to use dilutions giving less than ten plaques): For instance, if 63 and 57 plaques were counted at dilution 10^{-7}, the average of 60 should be multiplied by 10^7 (the dilution factor) and the by 10 (since 100 μl were used for infection). The resulting titer is 6×10^9 pfu/ml

3.7. Inactivation of rVV (see Notes 7 and 8)

Inactivation of viral replication generates a completely safe vector for research and clinical applications *(8, 13)*. Furthermore, this type of inactivated vector remains also "non-cytopathic" and most infected cells (at MOI below 10) can survive with no apparent negative effect. The technique used here is based on the random cross-linking of the genomic DNA of the virus with psoralen and long-wave UV. The resulting cross linking should be sufficient to abolish any possible replication but should also still enable early transcription which does not depend from the viral replication (as opposed to late viral protein).

1. Dilute the virus to a concentration of 5×10^8 PFU/ml in 1 ml of Hanks' Balanced Salts Solution (Invitrogen, Carlsbad, CA)
2. Add 10 μl of psoralen (100 μg/ml) and incubate 10 min at room temperature.
3. Transfer the solution in a 35-mm dish and place in the stratalinker (Stratagene, LaJolla, CA)
4. Remove the cover and irradiate with 365 nm UV light for 10 min (1.6 J)
5. Aliquot in 250 μl vials and freeze at −20°C

3.8. Expression of Recombinant Genes

The final verification of the construction of a recombinant virus is the characterization of the expression of the recombinant protein. In the case of the human CD80 costimulatory molecule, a simple antibody staining of the surface of infected cells will be sufficient to assess the presence of the new molecule.

1. Prepare a 6-well plate with 10^6 CV1 per well.

2. Perform the infection with control virus and CD80rVV inactivated viruses at MOI 10 (thus 10^7 pfu) following the same procedure as described before (leave a non-infected control)
3. Incubate from 6 to 24 h (for an easier detection, leave at least overnight)
4. Trypsinize the infected cells and transfer into FACS tube. Wash with PBS (add 2 ml of PBS, centrifuge 1,300 rpm for 2 min and decant supernatant)
5. Incubate for 30 min at 4°C with fluorescent-labelled specific human CD80 antibody or an isotype matched control reagent.
6. Wash twice with PBS and resuspend in 1 ml of fixing solution (4% paraformaldehyde) for 10 min.
7. Centrifuge 2 min at 1,500 rpm, decant the supernatant and resuspend in 300–500 µl PBS.
8. Acquire the cells with a FACS instrument.

Compared to the replicative counterpart virus, the characteristic expression from a virus inactivated with psoralen-UV usually displays a negative cell population resulting from infection with virus containing "cross-linked" gene thus unable to be transcribed. The frequency of "hits" in a gene is proportional to its length. However, as early transcription is no longer shut-off upon replication, the intensity of expression of an intact early gene is stronger as compared to its expression from a replicative recombinant virus (**Fig. 11.4**).

The use of CD80rVV during CTL stimulation in vitro and in vivo has demonstrated an increase capacity to activate CD8+ cells *(14, 10)*.

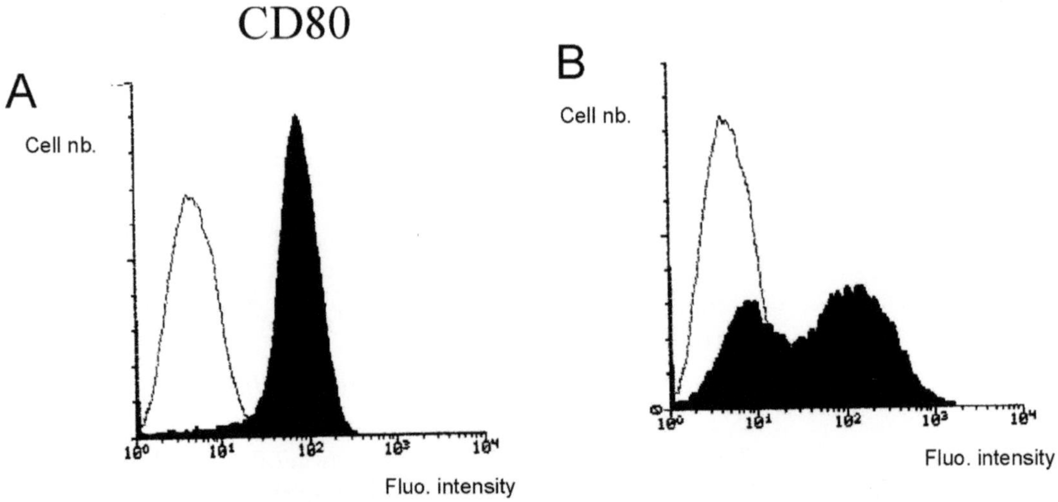

Fig. 11.4. Expression of human CD80 molecules from CD80rVV infected cells.
Expression of human CD80 as detected by specific antibody staining and FACS analysis. CV-1 cells were infected with (**A**) replicative and (**B**) inactivated virus (MOI 10). The *thin line* represents the staining of a negative control infection with wild-type virus.

4. Notes

1. This step is crucial for the efficiency of the following selection. Pay attention to the cells "quality" and perform, as described, a low moi infection.
2. After infection, cell-culture medium with only 5% FCS can be used instead of 10%.
3. Make sure your selective media is efficient by inhibiting the formation of plaques from a wild-type virus infection.
4. Make sure to select "real" replicative plaque especially at the very first cycle under pressure. Indeed, at this step there should be more than 10,000 wild-type virus for a single recombinant event. At this dose, even under pressure, wild-type virus can create some false small "starting plaques" resulting from the ECP of toxic effect of high-dose virus.
5. One can keep each cycle of selection at −20°C until the final selection in order to avoid to restart from the beginning in case of a problem with a cycle.
6. It is also recommended to select two plaques from each clone at the second isolation step without pressure in order to reach about a dozen of separate clones. Indeed each plaque selected after the first cycle without pressure contains a mixture of each possible forms of virus. Increasing the number of clones and passage increases the chance of isolating the recombinant form with no contamination with the other forms (intermediate and wt).
7. The level of inactivation of the virus may vary depending on the efficiency of the chemical solution and most of all, on the "age" of the UV bulbs. Thus it is recommended to evaluate the inactivation of virus by performing infections at different MOI. On CV-1 cells, correctly inactivated virus should display no major CPE before MOI of 15–20. But rec. virus should not be "over-inactivated" and verification of expression should also be performed in order to validate the procedure of inactivation.
8. Batches of virus are usually very stable at −70°C but one should avoid repeated freeze–thaw (more than three times) of each aliquot of inactivated virus as expression may strongly decrease.

References

1. Matzinger P. The danger model: a renewed sense of self. Science 2002;296: 301–5.
2. Goebel SJ, Johnson GP, Perkus ME, Davis SW, Winslow JP, Paoletti E. The complete DNA sequence of vaccinia virus. Virology 1990;179:247–63.
3. Johnson GP, Goebel SJ, Paoletti E. An update on the vaccinia virus genome. Virology 1993;196:381–401.
4. Moss B. Genetically engineered poxviruses for recombinant gene expression, vaccination, and safety. Proc Natl Acad Sci USA 1996;93:11341–8.

5. Moss B. Poxvirus vectors: cytoplasmic expression of transferred genes. Curr Opin Genet Dev 1993;3:86–90.
6. Bennink JR, Yewdell JW. Recombinant vaccinia viruses as vectors for studying T lymphocyte specificity and function. Curr Top Microbiol Immunol 1990;163:153–84.
7. Bennink JR, Yewdell JW, Smith GL, Moller C, Moss B. Recombinant vaccinia virus primes and stimulates influenza haemagglutinin-specific cytotoxic T cells. Nature 1984; 311:578–9.
8. Tsung K, Yim JH, Marti W, Buller RM, Norton JA. Gene expression and cytopathic effect of vaccinia virus inactivated by psoralen and long-wave UV light. J Virol 1996;70: 165–71.
9. Schlom J, Hodge JW. The diversity of T-cell co-stimulation in the induction of antitumor immunity. Immunol Rev 1999;170:73–84
10. Zajac P, Schutz A, Oertli D, et al. Enhanced generation of cytotoxic T lymphocytes using recombinant vaccinia virus expressing human tumor-associated antigens and B7 costimulatory molecules. Cancer Res 1998; 58:4567–71.
11. Kaufman HL. The role of poxviruses in tumor immunotherapy. Surgery 2003;134: 731–7.
12. Falkner FG, Moss B. Transient dominant selection of recombinant vaccinia viruses. J Virol 1990;64:3108–11.
13. Zajac P, Oertli D, Marti W, et al. Phase I/II clinical trial of a nonreplicative vaccinia virus expressing multiple HLA-A0201-restricted tumor-associated epitopes and costimulatory molecules in metastatic melanoma patients. Hum Gene Ther 2003;14:1497–510.
14. Marti WR, Zajac P, Spagnoli G, Heberer M, Oertli D. Nonreplicating recombinant vaccinia virus encoding human B-7 molecules elicits effective costimulation of naive and memory CD4+ T lymphocytes in vitro. Cell Immunol 1997;179:146–152.

INDEX

A

Affinity purification 139, 141, 145
Allogeneic B cells 22
Amicon stir cell 2, 6
Ampicillin 2, 4, 29, 31, 33
Anti-AsialoGM1 .... 17, 19, 23
Antibody production 22
Antibody-secreting cells 15
Antigen presenting cells (APCs) 9, 45, 50, 51, 52, 55, 66, 72, 73, 74, 82, 83, 84, 85, 86, 87, 88, 90, 91, 98, 99
Antigen specificity 21, 67, 90, 91
Anti-human IgE 19
Anti-human IgG1 19
Anti-human IgM 19, 22
Anti-PE MicroBeads 19
Aprotinin 140
Assay performance 123, 124, 131
ATP 3, 7
Autologous serum 70, 72, 73, 83, 89, 90

B

B95.8 cell line 71, 86
Band-pass 51
B cell helper assays 15–26
B cell proliferation 16, 21–22
B cells 15–26, 36, 40, 41, 42, 46, 59, 66, 86
BCLL 70, 71, 82, 85, 86
BioPlex 126
BirA substrate peptide 4
Bir A Enzyme 3, 7
Blue-native polyacrylamide gel electrophoresis (BN-PAGE) 135–149
Brefeldin A 11, 14
Brij96 .... 136, 137, 138, 139

C

C1R-CD1d 21
Calibration beads 37, 40, 45
Carboxyfluorescein Diacetate-Succinimidyl Ester (CFDA-SE) 17, 18, 21–22, 25, 52
CD1d 16
CD1d–Ig dimer 19
CD4 .... 3, 8, 9, 11, 16, 18, 20, 52, 55, 56, 57, 59, 65, 69, 120
CD4 T cells 60, 87
CD8 .... 3, 8, 9, 11, 12, 52, 54, 55, 56, 57, 59, 65, 120, 151–161
CD11a/b 11
CD19 59
CD20 .... 17, 20, 44
CD25 .... 13, 60, 65
CD27 .... 13, 17, 22
CD40 .... 15, 16, 18, 22
CD40L 16
CD44 .... 11, 13
CD45RO 13
CD56 59
CD80 151–161
CD161 17
Cell proliferation 16, 21–22, 85, 96, 99
Cluster 49, 59, 61, 129, 131, 135, 138, 139
Cluster analysis 49, 59, 131
Complementary DNA (cDNA) 4, 109, 110, 112, 113–114, 115, 152, 153, 154
Concanavalin A 123, 131
Coomassie Blue 5, 137, 142
Costimulation 15, 16, 159
Covalent coupling 122, 124–126
Cryopreservation 37, 43, 44, 48
CSK_4 .... 97, 100, 101, 102, 103
CTL 1, 2, 13, 151, 152
CTLA-4 65
CTL stimulation 160
CTNNB1 (catenin β-1) 29, 30, 31, 32
CV-1 cell line 152
Cyanine nucleic acid stains, 48, 54
Cyclosporine A 71, 86
Cytokine analysis 123, 129
Cytokine gene expression 107–117
Cytokines 1, 2, 3, 10, 11, 13, 14, 15, 16, 17, 18, 24, 36, 37, 39, 41, 42, 43, 44, 51, 66, 71, 82, 83, 87, 91, 95, 96, 107–117, 119–133

D

Data analysis 38, 41–42, 48, 49, 57–61, 104, 126
d-Biotin 3, 7
Diaflo PM10 .... 2, 6

Digitonin....136, 137, 138, 139, 140
Dilution factor....132, 159
DMSO....6, 18, 37, 38, 48, 71, 86, 103, 110, 141

E

EBV....71, 84, 86
ELISPOT....1, 95, 96
Emission wavelength....51
Epitope mapping....107–117

F

FACS buffer....3, 8, 11
Ferritin....136, 137, 143, 148
Ficoll....19, 20, 21, 42, 56, 70, 84, 108, 109, 123, 129
FITC....23, 50, 51, 52, 122, 125
Flow cytometry....3, 8, 10, 11, 13, 17, 20, 35–46, 47–62, 85
Fluorescence compensation....51, 52–54, 62
Fluorescence minus one controls....57
Fluorescence spillover....49, 53
Fluorochrome....4, 7, 47, 48, 49, 50, 51–52, 53, 54, 62, 117
Formaldehyde....3, 8, 11, 37, 38, 43, 44, 49, 160
Freezing....37, 54, 71, 75, 86–87, 89, 103, 158

G

αGalactosylCeramide....16, 18
Gene therapy....29
Geometric means....133
Glass homogeniser....6
Goat-anti-human IgG....19

H

Heat map....42, 129, 130, 133
Helper assay....15–26
HLA class I tetramers....2
Hly gene....29, 31, 33
^{3}H-Thymidine incorporation....21, 82, 100, 101
Human Ig ELISA....19
Hypotonic lysis buffer....141, 144
Hypoxanthine....153, 155, 156

I

Ig response....15, 19, 22, 25
IL-2....3, 10, 11, 17, 20, 37, 66, 70, 73, 76, 79, 82, 83, 84, 90, 91, 104, 119, 120, 122, 127, 128, 130, 131
IL-4....3, 10, 11, 17, 21, 40, 88, 119, 120, 122, 125, 126, 127, 128, 130
IL-6....37, 40, 120
IL-7....18, 51
IL-10....3, 10, 11, 37, 120, 122, 127, 128, 130
IL-13....17, 119, 120, 122, 124, 127, 130
IL-17....120, 122, 125, 126, 127, 130
IL-23....120, 122, 127, 130
Inclusion bodies....4, 5–6
INF-α....37
INF-γ....37
iNKT cell purification....23–24
Intracellular antigens....56–57
Intracellular cytokine staining....1, 3, 10
Inv gene....28, 29
IPTG....2, 4, 5, 29, 31, 33

J

JAK-STAT....36
Juvenile idiopathic arthritis....129, 130

K

Ki-67....49, 51, 55, 56

L

Leupeptin....3, 7, 140
Levy–Jennings plot....131
Limiting dilution....1, 20, 65, 69, 72, 75–81, 83, 87, 95–104
Limiting dilution assay (LDA)....1, 95–104
Linear regression....115, 116, 126, 128
LipofectamineTM....152
Long-pass....51
Luminex....123, 124, 126
Lysis buffer....111, 116, 140, 141, 143, 144, 145, 157

M

MACS®....21
Matrix effect....123, 124, 128, 129
Memory T-cell....108
Methanol....3, 37, 43, 44, 148
MHC class I....1–3, 4, 87, 88
β_2-microglobulin (β_2-m)....4, 5, 6
Microspheres....122, 123, 124–125, 126, 132
Mitomycin C....70, 82, 86
Mouse Ig ELISA....19
Multiplexed analysis....36
Multiplex immunoassay....119–133
Multi-protein complex....135, 136, 137, 139
Mycophenolic acid....152, 155

N

N-hydroxy-sulfosuccinimide....122, 124
NK1.1....17, 19, 23, 24, 25
NKT cells....23
Nonreplicating viral vector....151

O

Opti-Mem®....152, 154

P

P_3CSK_4....97, 98, 100, 101, 102
PE-Cy5....50, 51, 52
Pepstatin....3, 7
Peripheral blood mononuclear cells (PBMC)....19, 20, 21, 25, 35, 36, 37, 38, 40, 41, 43, 45, 48, 53, 55, 56, 59, 66, 70, 71, 72, 73, 74, 82, 83, 84, 85, 86, 88, 89, 90, 96, 97, 98, 99, 100, 103, 107, 110, 111, 129, 130
Pervanadate....136, 140, 143, 144
pET expression systems....2
Pfx DNA polymerase....29, 31
Phosphatase inhibitor....140, 141, 143
Phospho-antibodies....37, 44
Phosphoproteome....47
Plating efficiency....79–81
PMA....43
PMSF....2, 140
Poisson model....96, 97, 99, 102, 103, 104
Polychromatic flow cytometry (PFC)....47–62
Polystyrene microplates....22
Pouring blue-native gels....146–148
Poxviruses....152
Primary T cell culture....72–75
Propidium iodide....3, 8, 9, 48, 49, 55
Protease inhibitor....141
Psoralen-UV treatment....151

Q

Quantitative real-time PCR....112–113

R

Recombinant vaccinia virus....151–161
Refolding buffer....2, 6
Regulatory T cells (Tregs)....60, 121, 127
RNA....27–33, 109, 110–112, 117
RNAi....27–33
RNeasy Mini Kit....108, 110
RPMI 1640....3, 69, 96–97, 103, 108

S

Saponin....3, 11, 43, 44
SDS-PAGE....5, 138, 142, 148, 149
Sepharose....138, 139, 140, 141, 145
ShRNA....28, 29, 31, 32, 33
Single stained controls....57
SiRNA....27, 28, 29
Sodium fluoride....140
Sodium orthovanadate....140, 141
STAT....43
Stoichiometry....136, 143
Streptavidin....3, 4, 7, 8, 25, 123, 126, 138, 141, 145
Surface antigens....3, 8, 11, 13, 54

T

T cell clone....13, 15, 20–21, 65–91
T-cell epitopes....10
T cell priming....97–98
T-cell receptors (TCR)....4, 13, 16, 17, 23, 24, 65, 66, 67, 70, 91, 104, 135–149
TCRα....137, 138
TCR complexes....135–149
T-dependent B cell response....15
Terasaki....70, 76, 77, 78, 79, 80, 83, 90
Tetanus toxoid....66, 123, 129, 130, 131
Tetramers....2, 3, 4, 7, 8, 13, 26, 95
TM4 MeV....49
TO-PRO3....48, 54
Transkingdom RNA interference....27–33
Treeview....49, 61
Trip plasmid....28, 29, 30, 31, 33
Triton buffer....5

V

Vα14-Jα18....16
Vα24-Jα18....16
Vaccine formulations....151, 152
Vaccine vector....151
Vβ2....16
Vβ7....16
Vβ8.2....16
Vβ11....20, 67
Viability dyes....48
Viral genome....151, 155

X

Xanthine....153, 155, 156

Color Plates

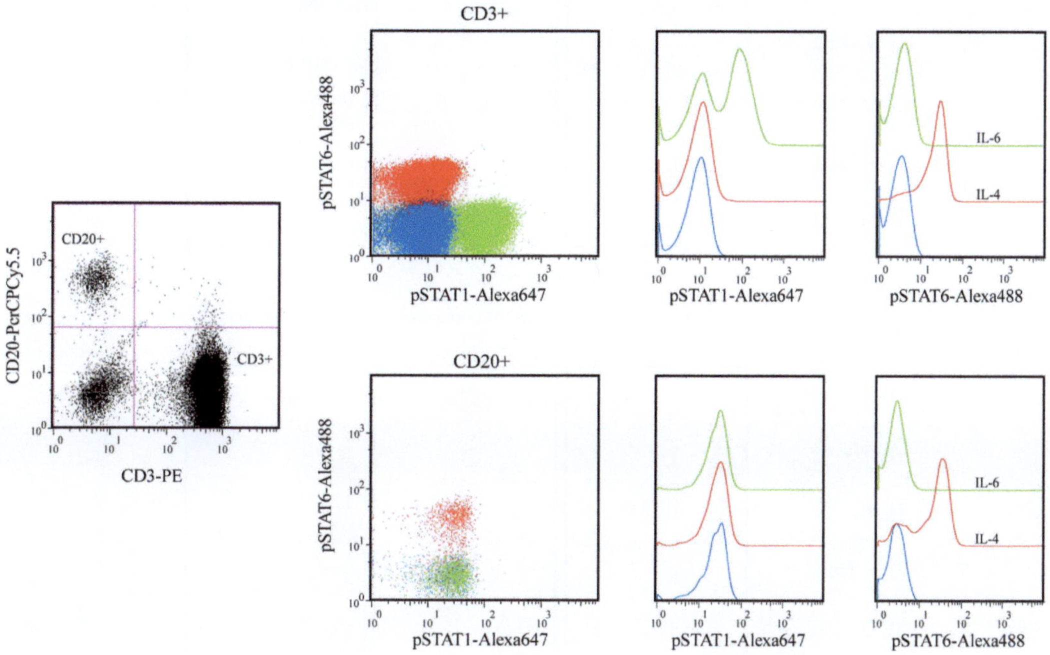

Color Plate 1. Multidimensional analysis of human PBMC stimulated with either IL-6 (*green*), IL-4 (*red*) or *left* untreated (*blue*).

Cells were fixed and permeabilized following protocol 3.2 and stained simultaneously with antibody cocktail A. *Top panels* show superimposed *dot plots* and *histograms* for T-cells ($CD3^+$), the *bottom panels* show B-cells ($CD20^+$). In overlays the induction of specific phosphorylation events are clearly identifiable. (*see* discussion on p. 41)

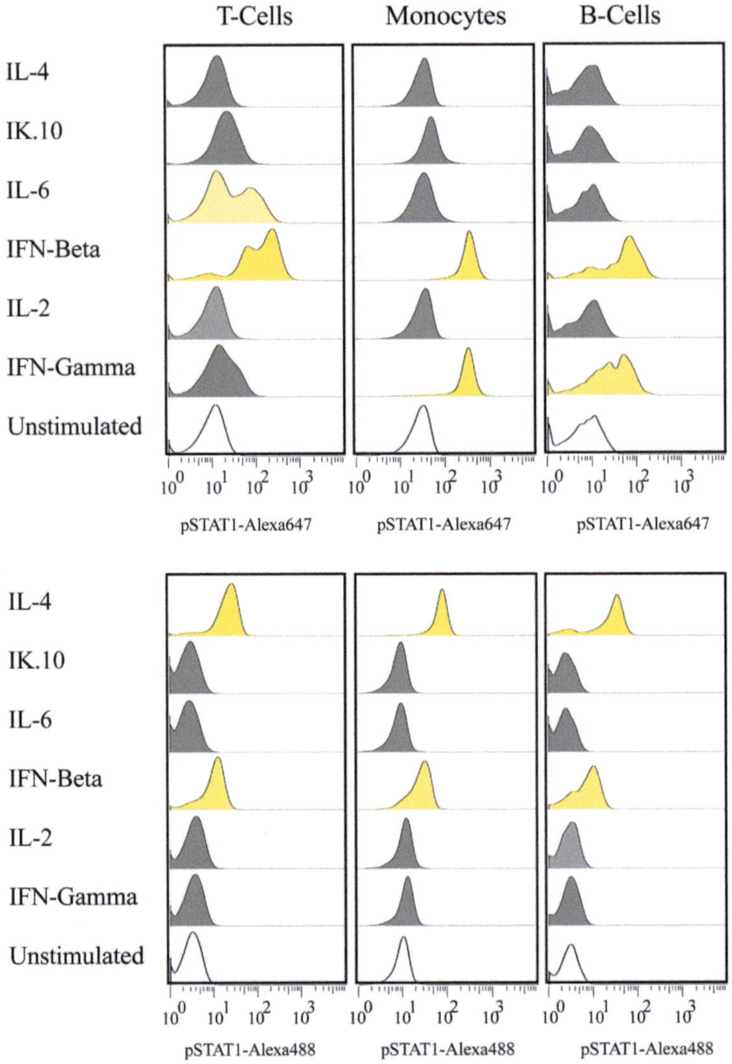

Color Plate 2. PBMC were stimulated with indicated cytokines, fixed and permeabilized according to protocol 3.2.
T-cells ($CD3^+$) and B-cells ($CD20^+$) were gated according to markers while monocytes were gated in *scatter plot*. *Open histograms* represent untreated cells, *filled histograms* stimulated cells. Induction of phosphorylation is clearly identifiable (*filled yellow histograms*). (*see* discussion on p. 41)

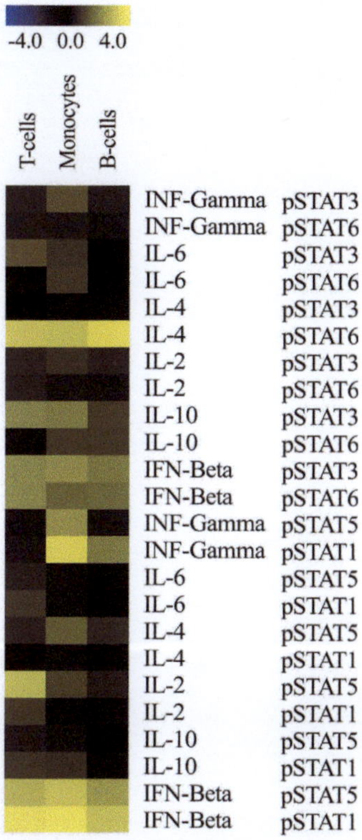

Color Plate 3. Visualization of the data generated by the FACS analysis following protocol 3.2. The columns represent the cell subsets, T-cells, B-cells, monocytes. Each row represents a cytokine stimulation stained with one of the antibody cocktails and subsequently analyzed for the indicated phosphoprotein. The color of each block represents the fold change ($\log_2$) in MFI in the channel corresponding to the analyzed phophorylated protein. (*see* discussion on p. 42)

MIX
Papier aus verantwortungsvollen Quellen
Paper from responsible sources
FSC® C105338
FSC
www.fsc.org

If you have any concerns about our products,
you can contact us on
ProductSafety@springernature.com

In case Publisher is established outside the EU,
the EU authorized representative is:
Springer Nature Customer Service Center GmbH
Europaplatz 3, 69115 Heidelberg, Germany

Printed by Libri Plureos GmbH
in Hamburg, Germany